RANGER URANIUM ENVIRONMENTAL INQUIRY

FIRST REPORT

Presiding Commissioner: Mr Justice R. W. Fox
Commissioner: Mr G. G. Kelleher
Commissioner: Professor C. B. Kerr

AUSTRALIAN GOVERNMENT PUBLISHING SERVICE
CANBERRA 1976

ISBN 0 642 02260 7

Printed by Authority by the Government Printer of Australia

RANGER URANIUM ENVIRONMENTAL INQUIRY

28 October 1976

In accordance with the *Environment Protection (Impact of Proposals) Act* 1974 and the terms of our appointment we have the honour to present this our First Report.

R. W. Fox
Presiding Commissioner

G. G. Kelleher
Commissioner

C. B. Kerr
Commissioner

The Honourable K. E. Newman, M.P.
Minister of State for Environment,
Housing and Community Development
Parliament House, Canberra, A.C.T. 2600

PREFACE

Uranium is a very special metal: it contains fissile atoms. It is used in nuclear reactors to produce heat which converts water to steam. The steam drives turbines to generate electricity. There are substantial deposits of uranium ore at the Ranger site in the Northern Territory. The Australian Atomic Energy Commission and Ranger Uranium Mines Pty Ltd have a proposal to mine and mill the uranium. This Commission was established to inquire into the environmental aspects of that proposal. The uranium oxide which comes from the mill is not to be used in Australia, but is to be exported to countries which produce electricity from nuclear reactors. It was submitted to the Commission that there are serious risks and disadvantages associated with the various operations of the nuclear power industry, from mining and milling the uranium to disposal of the radioactive wastes from the nuclear reactors. The matters mainly relied on were releases of radioactivity, particularly accidental releases, the possibility of accidental nuclear explosion, the high-level radioactivity of wastes and the possibility of terrorist use of the plutonium produced in reactors. It was submitted also that extension of the nuclear power industry involved increased risks of nuclear war, flowing from the availability of plutonium, or enriched uranium, for atom bombs. It was submitted that because of all those considerations, and others as well, Australia should not sell its uranium, or mine it. Those objections are examined in this First Report. The more local environmental aspects which relate particularly to the Ranger proposal will be dealt with in the Second Report. The wider matters mentioned are here considered in the context of the world energy situation, the need and demand for uranium, the amount likely to be earned by its sale, and other economic considerations. This Report also looks at the question of alternative sources of energy and the possibility of conserving energy resources.

After this Report was written, the Sixth Report of the British Royal Commission on Environmental Pollution was published in the U.K. Its subject is Nuclear Power and the Environment, and it canvasses many, but not all, of the issues which we have considered. We comment on it in a Postscript at the end of this Report.

CONTENTS

LIST OF PLATES

LIST OF TABLES

1 INTRODUCTION

By instrument published in the *Australian Government Gazette* on 16 July 1975, the Prime Minister and Minister of State for Environment, The Honourable E. G. Whitlam, Q.C., acting in the latter capacity, directed in pursuance of section 11 (1) of the *Environment Protection (Impact of Proposals) Act* 1974 that an inquiry be conducted in relation to the proposal for the development by the Australian Atomic Energy Commission in association with Ranger Uranium Mines Pty Ltd of uranium deposits in the Northern Territory of Australia. By a further instrument published on the same day, the Honourable Russell Walter Fox, Graeme George Kelleher Esquire and Charles Baldwin Kerr Esquire were appointed a Commission to conduct the Inquiry, and the Honourable Russell Walter Fox was appointed to preside at the Inquiry. The Commissioners are respectively the Senior Judge of the Australian Capital Territory Supreme Court, a civil engineer and Hearings Commissioner, and the Professor of Preventive and Social Medicine at the University of Sydney.

The Commission is required to inquire:

in respect of all the environmental aspects of:

(a) the formulation of proposals;
(b) the carrying out of works and other projects;
(c) the negotiation, operation and enforcement of agreements and arrangements;
(d) the making of, or the participation in the making of, decisions and recommendations; and
(e) the incurring of expenditure,

by, or on behalf of, the Australian Government and the Australian Atomic Energy Commission and other authorities of Australia for and in relation to the development by the Australian Atomic Energy Commission in association with Ranger Uranium Mines Pty Ltd of uranium deposits in the Northern Territory of Australia.

'Environment' is defined by the Act to include 'all aspects of the surroundings of man, whether affecting him as an individual or in his social groupings'.

Procedure of the Inquiry

The Act and Regulations thereunder control the procedure of inquiries instituted under the Act. Subject to specified exceptions, an inquiry is to be held in public, and evidence in the inquiry is required to be taken in public on oath or affirmation. Each witness, with the leave of the Commission, is entitled to cross-examine any other witness. The Commission accepted and acted on the view that, subject to the operation of the exceptions, it was not entitled to elicit evidence in private, but was confined to the evidence given at public hearings. The specified exceptions referred to enable the Commission to receive evidence in private when satisfied that it is desirable to do so in the public interest 'by reason of the confidential nature of any evidence or matter or for any other reason', and to receive a verified statement from a person when his attendance as a witness 'would cause serious hardship to the person'. In the event, only one witness was allowed to give evidence in private: he was a government officer who dealt with some aspects of the subject of terrorism. A number of statutory declarations were received under the other exception mentioned, and their contents were made public.

The administrative procedures prescribed under the Act provide that the Minister may require the preparation, by proponents, of a 'draft environmental impact statement' and also provide for that document being made available to the public for comment. In the present case, the managing company, Ranger Uranium Mines Pty Ltd, had prepared a document called 'Environmental Impact Statement' before 20 June 1975, when the administrative procedures came into effect. This document was treated as if it had been prepared in response to the requirements of the administrative procedures, although it did not in all respects comply with those requirements. By notices published in newspapers throughout Australia in July 1975 the public was notified that that document was available for inspection or purchase, and the prescribed period of not less than 28 days was allowed for submissions about it to be sent in. The notice was in fact issued in the name of a Commission constituted earlier to inquire into the same proposal, but which was superseded by the present Commission, with fresh terms of reference, by the instruments already referred to. To save time and expense the present Commission did not issue fresh notices. Many submissions were received, most of which opposed the proposal set out in the Environmental Impact Statement (EIS) on one or more of a wide variety of grounds.

Scope of the Inquiry

Many were opposed to the proposal on the ground that the use of uranium in the nuclear power industry carried with it risks and dangers of such a nature and magnitude that Australia should not export it, or mine it at all. This aspect was not dealt with in the EIS, although environmental considerations clearly were involved. The Commission informed the proponents that it held the tentative view that it could and should inquire into the matters mentioned, but it invited submissions as to the correctness of that course. No contrary submissions having been received, the Inquiry proceeded to take evidence and hear submissions in relation to those matters, as well as those of more localised impact. The Commission's remit was not to inquire into Australian uranium mining activities as a whole. It had, however, to examine the wide grounds of objection in relation to the particular proposal into which it had been appointed to inquire.

It is well accepted that a cost-benefit analysis is a part of any environmental inquiry or investigation. It was therefore in any event necessary to consider the present and future market prospects of uranium, but the wider aspect mentioned made more direct the necessity of considering the need for uranium as a source of energy, and hence world and national energy resources.

Parliamentary Inquiries on related matters

Before public hearings commenced, Mr John Kerin, M.P., Chairman of a sub-committee appointed by the House of Representatives Standing Committee on Environment and Conservation to inquire into the environmental implications of the mining, processing and use of uranium in Australia, had discussions with members of the Commission about the ambit of its inquiry. We understand that, because the Commission was dealing with similar issues, the Committee deferred its inquiry in order to avoid duplication.

More recently, we have had the benefit of discussions with reference to solar energy with Senator Thomas, now Chairman of the Senate Standing Committee on National Resources, and Mr Edwards, Secretary of that Committee.

Advisors to the Commission

Under the Act the Minister may appoint persons to advise the Commission, and in September 1975 the Minister exercised this power by appointing the undermentioned persons, whose relevant areas of specialisation are shown beside their names:

Mr Clifford Christian—Ecology and Land Use
Mr Chester Gray—Nuclear Technology and Chemical and Metallurgical Engineering
Mr William Gray—Aboriginal Welfare
Dr Barry Hart—Water and Water Quality
Mr William Holder (resigned 15 January 1976)—International Law and Relations
Mr John Lake (resigned 3 November 1975)—Fresh Water Biology
Professor Gregory McColl—Economics
Dr Ronald Rosen—Radiation Protection
Professor Donald Greig, appointed vice Mr William Holder, with effect from 28 April 1976—International Law and Relations

These advisors were available on a part-time basis, except that Mr Chester Gray was available full-time until 10 September this year.

The presentation of evidence

Public hearings began in Sydney on 9 September 1975 and were held in Sydney, Darwin, Mudginberri (near the proposed mine site), Gove, Brisbane, Adelaide and Melbourne. Arrangements were made for witnesses from Tasmania, Western Australia and the Australian Capital Territory to give evidence at one or other of the places mentioned, either in person or by statutory declaration. A total of 281 persons gave evidence, and 354 documentary exhibits were received in evidence, many of which were books or papers by specialists. The transcript of evidence occupied 12 575 pages.

In July 1976 the Commission made a two-week visit to England and the Federal Republic of Germany, where it saw nuclear plant and facilities in commercial operation, and associated research laboratories.

Public hearings concluded in Sydney on 12 August 1976.

A list of the names of the persons who gave evidence is at Appendix A. In a number of cases they spoke for or represented organisations, which have not been separately named.

The Act and subordinate legislation do not make provision for the recognition of any parties, other than the proponents. The Commission, nevertheless, found it convenient to give a degree of independent recognition to some companies or organisations which had or manifested a particular interest in the subject of the Inquiry. The undermentioned parties were recognised, and they accepted the Commission's invitation to submit final written submissions (and replies to the submissions of the others) and to be represented on the closing days of the inquiry to answer queries from the Commission:

Australian Atomic Energy Commission
Ranger Uranium Mines Pty Ltd
Australian Conservation Foundation
Conservation Council of South Australia Incorporated

Friends of the Earth
Northern Land Council
Northern Pastoral Services Ltd
Oenpelli Council
Pancontinental Mining Ltd

This Report

The Commission is required to make findings and recommendations and to report them to the Minister. It is further required, after so reporting, to make public those findings and recommendations. The style and content of this Report have been influenced by this latter requirement.

Until a late stage, it was the intention of the Commission to have one principal report. In June 1976 the Commission came to the conclusion that it should consider the possibility of dealing first, in a separate report, with the ground of objection to the effect that Australia should not mine or sell uranium at all.

There are a number of disadvantages, from the point of view of others as well as the Commission, in our following that course. In the circumstances, we have decided that the desirability of an early statement of the facts, and expression of our views relative to that aspect, outweighs these disadvantages. We have divided the subject in the way mentioned, and this Report, to be known as our First Report, will deal with the ground of objection referred to.

It has to be recognised that this Report cannot provide a total evaluation of the question whether Australia should or should not supply uranium to other countries. To grapple with that question involves looking more closely at other Australian mining proposals, making if possible a total cost-benefit appreciation of the position, and examining a wide range of considerations affecting Australia's external relations. These are largely beyond the ambit of the Inquiry. What we can do, and what we attempt, is to examine the facts relating to the hazards and problems associated with the handling and use of uranium and to evaluate those hazards and problems so far as the scope of the Inquiry permits. On the basis of that evaluation, we make some findings and recommendations.

We had hoped to be able to deal with relevant matters in a short and simple manner in one document, leaving the longer more technical discussions for an accompanying document. Our decision to bring in a First Report so soon has meant that we are not able to adopt that course, because the omission now of technical discussions would, we believe, leave many people at a disadvantage as against people claiming superior technical or scientific knowledge. The result is that, for many, parts of this Report will make heavy reading.

The Commission's Second Report

We will proceed immediately to finalise and present our Second Report, which will deal with the remaining issues. We wish to emphasise that the fact that we proceed to consider the Second Report does not in any way involve anticipation by us of decisions to be taken as a result of this Report.

The Second Report will deal with a complex of serious issues respecting the particular mining proposal. Most of the evidence has been concerned with those issues.

They include the matters with which the EIS deals. We do not at this stage express any view on them. It is common ground, we believe, that some at least are highly sensitive and difficult of resolution. These include the position of the aboriginal people who oppose mining in the area, and some of whom wish to be treated as owners of the land sought to be mined. There are a number of serious problems which will arise respecting the culture and future welfare of the aboriginal people if mining in the area goes ahead, and these have to be considered.

A major National Park is proposed within what has been called the 'uranium province', which lies roughly between the South Alligator and East Alligator Rivers, and we have been asked to suggest where its boundaries should lie, having in mind proposed and possible mining operations. Not only is there the present development proposal by Ranger Uranium Mines Pty Ltd and the Atomic Energy Commission, but there are other nearby deposits which Ranger Uranium Mines Pty Ltd may wish to develop in the future, and both Pancontinental Mining Ltd and Noranda Australia Ltd have deposits not far away which they wish to develop. The boundaries of the National Park, as presently proposed, include part of the deposit of Pancontinental Mining Ltd and the whole of the deposit of Noranda Australia Ltd, but pass just outside the Ranger Uranium Mines Pty Ltd area. There is an important question of the impact of mining operations on the National Park. It is intended that there be a regional centre to accommodate people associated with the mine and their families, and visitors, and if the three mines mentioned were to be developed this would involve an estimated population of up to 15 000 people. Consideration has to be given to the effects of such a township. There is also the position of the owners of a nearby large grazing property, whose fear is that it may be affected by effluents from the mine, including airborne and waterborne radioactive material. Fears have been expressed as to the effect on wetlands and wildlife habitats of the mining operations. If these considerations, taken singly or together, were not to be regarded as sufficient to stand in the way of mining, it would still be our duty to recommend courses designed to obviate or minimise any undesirable or deleterious consequences.

Relative importance of findings of fact

This First Report of necessity deals with some very broad issues, with respect to which different minds can quite readily come to different conclusions. Having regard to this, and the limitations on the ambit of the Inquiry, the view might be taken that, in relation to those issues, our findings of fact will prove more valuable than our final recommendations. In any event, we wish to stress the relative importance of those findings. As we understand the intended operation of the Act, it is that members of the public (as well as the Minister) be provided with findings of fact as determined by an independent tribunal, so that they can form their own opinions and, if necessary, influence parliamentary and government action. They can support the particular recommendations or not, as they feel appropriate, but we hope they will at least have an account of the facts upon which they can rely.

Witnesses

In considering the evidence, we have found that many wildly exaggerated statements are made about the risks and dangers of nuclear energy production by those opposed to it. What has surprised us more is a lack of objectivity in not a few of those in favour of it, including distinguished scientists. It seems that the subject is one very apt to arouse strong emotions, both in opponents and proponents. There is abundant evidence before us to show that scientists, engineers and administrators involved in the business of producing nuclear energy have at times painted excessively optimistic pictures of the safety and performance, projected or past, of various aspects of nuclear production. There are not a few scientists, including distinguished nuclear scientists, who are flatly opposed to the further development of nuclear energy, and who present facts and views opposed to those of others of equal eminence.

We note that a few of the government officials who appeared before us showed reluctance in communicating matters of importance to the Commission. It seemed to us that the objectives and working of the Environment Protection (Impact of Proposals) Act may not be clearly understood in some government departments.

Ultimately, when the matters of fact are resolved, many of the questions which arise are social and ethical ones. We agree strongly with the view, repeatedly put to us by opponents of nuclear development, that, given a sufficient understanding of the science and technology involved, the final decisions should rest with the ordinary man and not be regarded as the preserve of any group of scientists or experts, however distinguished.

A few of the publicists for nuclear development characterise their opponents as lobbyists or dissidents, or worse. We would wish to make it quite plain that before us the opposition has come from a wide cross-section of the general community, and we would not be prepared to conclude that their motives and methods are any less worthy or proper, or intelligently conceived, than, in general, are those of the supporters of nuclear development.

British Inquiry on related matters

The British Royal Commission on Environmental Pollution undertook in early 1974 an inquiry into nuclear power, with particular reference to its impact on the British environment, and it is due to report soon. It will undoubtedly deal with some matters also the subject of this Report. This is the only public inquiry of which we are aware, apart from the present, which has endeavoured to look at the whole industry from an environmental point of view. While Britain is a nuclear weapons power and has a substantial nuclear power industry, Australia does not have, nor, so far as we are aware, does it presently intend to have, nuclear weapons, and no plans have been announced for it to have a nuclear power industry. In fact, it is very fortunately placed so far as concerns coal, the principal alternative fuel for electricity generation. It nevertheless has substantial deposits of uranium of high grade, which can be won relatively cheaply. As it has uranium for export but no dependence upon a domestic nuclear industry, it is able to look upon the nuclear energy scene from a viewpoint denied other developed countries.

The need for regular review

We understand that it is unlikely that there can be production from any of the mines we have mentioned before about 1980. The expected lives of the mines can be taken as of the order of 30 to 40 years, but may be longer. When considering the supply of uranium, this Report concentrates more particularly on the period 1980-2000. In the ordinary course of uranium trading, contracts would be entered into several years, at least, before the intended date of supply.

To quite an extent, it becomes necessary to forecast the future on the basis of present knowledge. This has been done by a considerable number of specialists in the various fields, and we have the published views of many of them. What has to be explained, we think, is that knowledge concerning matters nuclear is expanding at a very great rate. It is a new area, and one about which much remains to be learned. We therefore suggest, at the outset, that conclusions now reached and decisions now arrived at should, in general, be reviewed within a few years, and periodically thereafter.

Legislative control

Uranium mining in Australia has, since 1946, been subject to special controls under Commonwealth legislation. Uranium in the Territories is vested in the Commonwealth (*Atomic Energy Act* 1953, s. 35). Under s. 38 of that Act power is given to make regulations giving wide control to the Government in respect of all uranium, wherever situate, but the power, in its application to the States, is only to be exercised for defence purposes. No regulations under that power are in operation. Where in this Report we refer to the mining of uranium in Australia, without differentiation between States and Territories, we do so because of the form in which the arguments were presented to us. No discourtesy to the Legislatures or Governments of any State or Territory is intended. So far as we have any views with regard to where legislative power should reside, we will express them in this or the ensuing Report.

Acknowledgments

We wish to express our warm appreciation of the conscientious assistance which our advisors have given us. The Report does not necessarily represent the views of any of them. The Commission takes responsibility for the whole of it.

The unreasonably short time the Commission set for preparation of this First Report made very heavy demands on its staff, who responded marvellously. We are very grateful for their willingness to work very long hours over a substantial period of time. In particular, we mention at this time Mr John Rames, Secretary of the Commission, Dr Hugh Saddler, Research Officer, Dr Gordon Aitchison, who assisted in the writing of Chapters 3 and 10, and Mr Robert Lehane, who joined us in July this year as Editorial Manager.

Scheme of the Report

The scheme of this Report is to deal in a series of chapters with the principal relevant topics. In Chapter 15 we bring together some considerations concerning world energy resources and needs, and concerning the related matter of proliferation, and make suggestions as to future action in relation to both those topics. We discuss our conclusions in Chaper 16, and we then state our principal findings and recommendations.

Figure 1

WHERE THE URANIUM IS

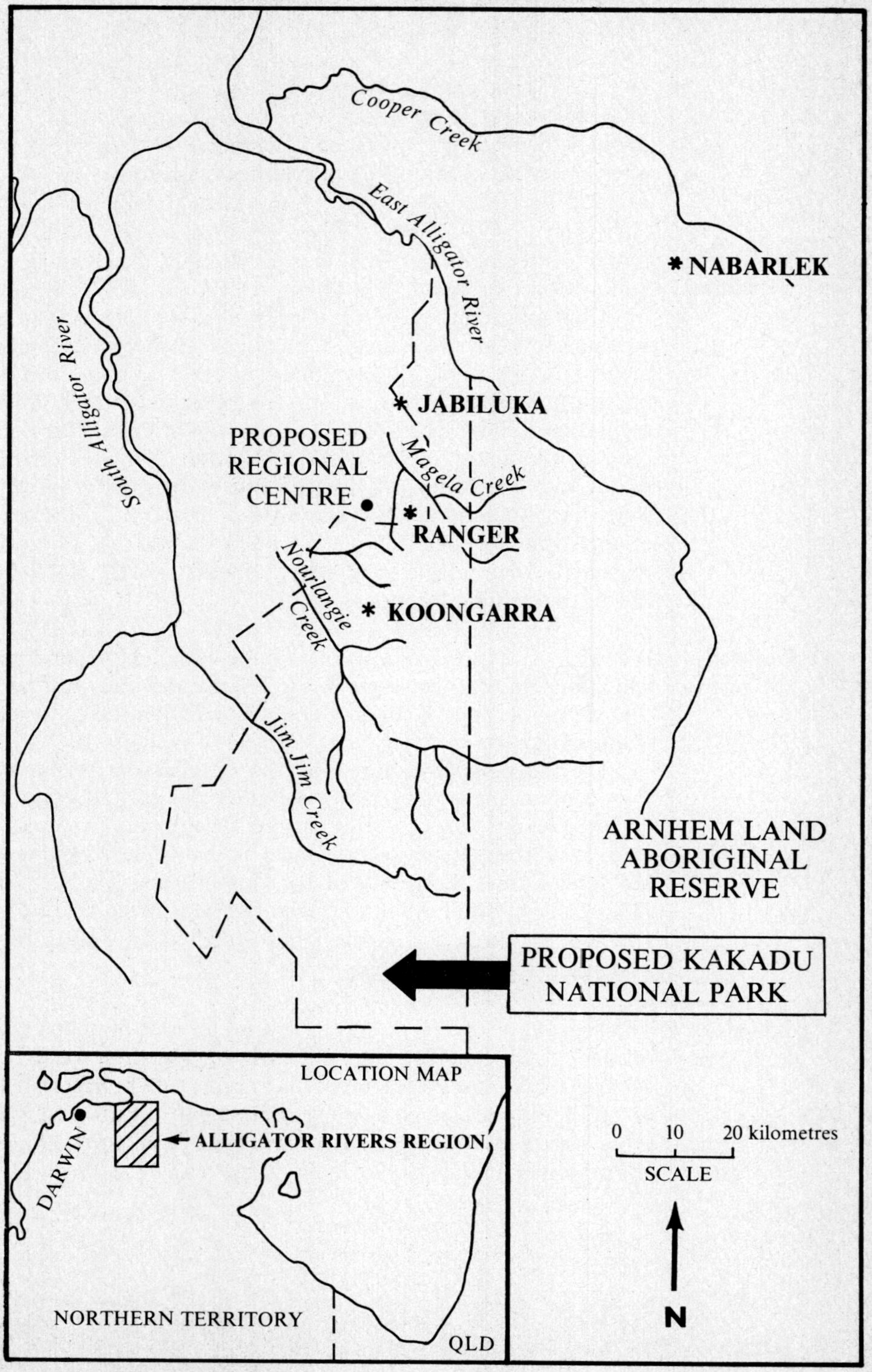

* Uranium deposits
– – – Proposed national park boundary

2 THE RANGER PROPOSAL

Exploitation of the Ranger uranium deposits, situated near the Arnhem Land Aboriginal Reserve about 220 kilometres east of Darwin, is envisaged as a joint venture involving the Commonwealth Government, Peko Mines Ltd and Electrolytic Zinc Company of Australasia Ltd. The plan is to mine the ore and convert it to yellowcake, which is the concentrate produced in the first processing step between uranium ore and nuclear fuel; it contains more than 90 per cent U_3O_8 (a form of uranium oxide). The yellowcake is then to be exported to countries wanting it for nuclear power production.

The managing company for the project, Ranger Uranium Mines Pty Ltd, claims that reasonably assured resources at the site, recoverable at a cost of up to US$15 per pound of U_3O_8, total 85 000 tonnes of uranium. The company has put forward an annual production target of 3000 tonnes of U_3O_8 in yellowcake rising to 6000 tonnes if markets for that amount are available.

According to a memorandum of understanding between the co-venturers, dated 28 October 1975, the Australian Atomic Energy Commission (AAEC) will provide 72½ per cent of the capital required for the project and Peko and Electrolytic Zinc 13¾ per cent each. Two directors will be appointed to Ranger Uranium Mines Pty Ltd by the AAEC and one each by the two companies. The memorandum gives the AAEC responsibility for all future sales of yellowcake from the mine.

The Commission was told that the present Government has examined a possible alternative arrangement under which the Australian Industry Development Corporation might take over the Commonwealth's financial role. It was told that the Government did not intend to carry further its consideration of this arrangement, or of any other possible arrangement, until this Commission had reported.

Three other large uranium deposits have been found not far from the Ranger site: Koongarra, about 24 kilometres to the south, the exploration licence for which is held by Noranda Australia Ltd; Jabiluka, about the same distance to the north, the exploration licence for which is held by Pancontinental Mining Ltd; and Nabarlek, some 65 kilometres to the north-east, the exploration licence for which is held by Queensland Mines Ltd. Ranger and the three deposits nearby account for the bulk of Australia's presently known uranium resources.

They are located in a region (see Figure 1) that is in many ways remarkable. It is largely undeveloped. The Aboriginal population still maintains elements of its traditional way of life, and the spectacular Arnhem Land escarpment contains many of the best surviving examples of Aboriginal rock art. Numerous rich archaeological sites provide evidence of continuous human habitation for tens of thousands of years. The area is noted for its abundant wildlife, particularly the waterbirds that gather on its floodplains. Much of its vegetation is diverse and interesting.

24647/76—2

Drilling at the Ranger site has defined two large ore-bodies, about 2 kilometres apart. One is said to be about 400 metres wide and 180 metres deep; the company plans to mine it first, using open-cut methods. This ore-body is estimated to contain about 20 million tonnes of ore. Tests indicate that it averages 0.25 per cent uranium oxide, so about 450 tonnes will have to be mined and processed to produce each tonne of yellowcake.

The second ore-body is about the same size. It is in two parts, one on top of the other. The upper portion is said to extend from the surface to a depth of about 200 metres, and the lower part to go down to about twice that depth. It is proposed that the top portion be open-cut mined. The rest might be mined by underground methods; this remains to be determined.

How long the mine will be able to remain in production depends on many factors, including production rates and the results of further exploratory drilling. The mine's lifetime was estimated in evidence by the company to be between twenty and thirty years, depending on the rate of production. Further ore bodies in the immediate vicinity may be developed, resulting in a longer life for the mine.

The first definite indication that the Ranger area contained rich uranium reserves came in June 1970 when an aerial survey detected radiation anomalies there. The survey was being conducted as part of a joint exploration program, by Peko Mines Ltd and Electrolytic Zinc Company of Australasia Ltd, which had begun three years earlier. Follow-up investigations on the ground confirmed the find, and in June 1971 the two companies formed Ranger Uranium Mines Pty Ltd, to manage development of the deposit.

The Commission was told that the Ranger company had spent about $5¾ million since the discovery on detailed assessment of the reserves and planning for their development. A workforce of up to 60 has been involved in these preliminary activities. It has been accommodated at the settlement of Jabiru near the proposed mine site.

The company anticipates employing up to 600 people during the construction phase for an initial production rate of 3000 tonnes a year, or 1000 if it is decided to develop a 6000 tonnes a year capacity immediately. The construction phase would last at least three years, and the workforce would be housed at a construction camp on the site. The number of people employed would fall when the mine and yellowcake mill began production. The company's estimates of the size of the operating workforce are 250 at a production rate of 3000 tonnes a year or 400 for the higher rate.

The company hopes that a proposed new regional centre, which would be 8 kilometres west of the Ranger deposit, will be built during the construction phase. If this happens, it plans to build permanent housing for its workers there.

The area the company expects to utilise if mining proceeds is about 900 hectares—150 hectares for roads, camps, a temporary township and the airstrip, and the rest for mining activities. These activities will make big changes to the landscape. Material other than ore dug out during mining will be placed on a waste dump, which will cover more than 100 hectares. The slurry remaining after the ore passes through the yellowcake mill will be piped into a tailings dam, also covering more than 100 hectares. The company's plans provide for the eventual revegetation of the whole area, except for the mine pits (which will become lakes) the airstrip, and some roads.

It is proposed that liquids used in the milling operation should be recycled so that none is intentionally released to the environment. However, some seepage from the tailings dam will occur. Some of this, together with contaminated runoff from rainfall on the mine and mill area, will initially be contained in a retention system. The intention is to release it to the environment at times of high river flow in later years, on the 'dilute and disperse' principle. A sulphuric acid plant proposed for the site would release some sulphur dioxide to the environment.

It is important to bear in mind that it will be several years before the mine can commence operations. If approval were given to go ahead in, say, early 1977, yellowcake would probably not be exported before 1981. The evidence is to the effect that the Ranger mine is at a more advanced stage of readiness to produce than any other uranium mine in Australia, with the exception of the relatively small mine at Mary Kathleen in Queensland which is in production. The Mary Kathleen mine produced uranium from 1958 to 1963, but then closed down for want of a market. It re-commenced production this year, and now is satisfying contracts at the rate of about 1000 tonnes per year. The period of time principally under consideration in this First Report, when considering the possibility of export of Australian uranium from new mines, is from about 1980 to about 2000.

Plate 1. Mary Kathleen uranium mine near Mount Isa, Queensland. The open-cut mine is in the foreground, a waste dump can be seen behind it, the yellowcake mill is in the middle ground on the left and the tailings dam is in the background. (Photo by courtesy of Mary Kathleen Uranium Ltd.)

Plate 2. This photo, of a uranium mine and yellowcake mill at l'Ecarpiere, France, shows the mill in the foreground and the edge of the tailings dam on the right. Behind the mill can be seen the small open-cut section of the mine and the tops of two shafts through which ore is extracted from the underground section. (Photo by courtesy of Commissariat a l'Energie Atomique.)

3 THE BASICS OF NUCLEAR POWER

Uranium is a totally different type of fuel from all its predecessors and present-day competitors. Whatever their form—solid, liquid or gas—the other fuels are carbon compounds that react with oxygen when they burn, producing heat and carbon dioxide. The chemical elements present before combustion still exist in the same quantities afterwards.

Uranium gives off energy when atoms of the element split apart (i.e. undergo fission). The uranium is replaced by other elements. The production of energy by fission involves the conversion of a small part of the original mass into energy; because of this, the combined mass of the products of fission is slightly less than the mass of the original uranium.

Induced fission was first observed in the laboratory in 1938. The first practical application of the process, the atomic bomb, was followed after World War II by the development of electricity generating plants making use of the heat energy given off in controlled fission reactions.

In line with the basic difference in the way the energy is produced, the effects on the environment of conventional and nuclear electricity production are very different. These are discussed in Chapters 10 and 11.

In this Chapter, the main features of nuclear power production are briefly outlined.

The atomic nucleus

An atom is the smallest particle into which an element can be divided chemically. It consists of a nucleus of protons and neutrons surrounded by a cloud of electrons. The number of protons in the nucleus determines what element the atom represents; for example, hydrogen has one proton and uranium ninety-two. Protons and neutrons are more than 1800 times as massive as electrons, so almost the entire mass of an atom is concentrated in its nucleus.

Protons and electrons carry equal electrical charges—protons a positive charge and electrons a negative one. Since an atom is electrically neutral, it must have the same number of protons and electrons.

Neutrons, as their name implies, are electrically neutral. Different atoms of the same element can have different numbers of neutrons; such atoms are known as isotopes of the element. For example, some naturally-occurring uranium atoms have 143 neutrons while most have three more. These isotopes are the nuclides, or 'nuclear species', uranium-235 and uranium-238, also written as ${}^{235}_{92}U$ and ${}^{238}_{92}U$. The top number is the sum of protons and neutrons in the atom, and is known as the mass number. Below it is the atomic number, the number of protons in the atom.

In nature, unless there is some fractionating influence, different isotopes of any element always occur in the same proportion. Most natural hydrogen, for example, has no neutrons accompanying the single proton in its nucleus; it is hydrogen-1 or ${}^{1}_{1}H$. But one atom in every 6000 is hydrogen-2 (${}^{2}_{1}H$), known as deuterium which has one neutron and one proton in its nucleus. Uranium-238 is the predominant isotope in all uranium deposits, but one atom in 140 is the type that splits apart in atomic bombs and nuclear reactors —uranium-235.

Radioactivity

All nuclides with more than 83 protons, and some with fewer, are unstable. All atoms in a sample of an unstable nuclide ultimately change spontaneously into other nuclides (decay products) by the process of radioactive decay.

Some emit alpha particles, which consist of two protons and two neutrons, when they decay. These alpha emitters become atoms of the element two down the scale of atomic number; for example, uranium-238 ($^{238}_{92}U$) becomes thorium-234 ($^{234}_{90}Th$). Other atoms shoot out beta particles in the form of negatively charged electrons. Neutrons in the nucleus thereby turn into protons, so the result of such a beta emission is an increase of one in atomic number. An atom of thorium-234 ($^{234}_{90}Th$), for example, becomes an atom of protactinium-234 ($^{234}_{91}Pa$). There are also man-made nuclides which emit on decay beta particles in the form of positrons (essentially positively charged electrons), and so decrease their atomic number by one. In such cases protons in the nucleus turn into neutrons.

The other form of radioactive emission is gamma radiation. This resembles x-radiation in that it is not particulate, so no change occurs in proton and neutron numbers when a nucleus emits it. Gamma rays often accompany alpha or beta emissions.

Uranium-238 decays through a series of 14 alpha and beta decay steps to a non-radioactive isotope of lead. Uranium-235 also decays, through a different sequence, to another stable lead isotope. However, this does not mean that the world's uranium reserves will soon turn themselves into lead. Uranium-235 is a very stable radioactive isotope; it has a half-life of 713 million years, which means that half of any sample will decay in that amount of time. Uranium-238 has an even longer half-life, 4510 million years, which appears to be about the length of time since the earth and the rest of the solar system were formed. This means we still have half the uranium-238 that was present when the earth formed. Because of the difference in half-lives, the ratio of uranium-235 to uranium-238 has very gradually decreased over the earth's history.

Radioactive nuclides have half-lives ranging from fractions of millionths of a second to thousands of millions of years. Short-lived nuclides emit radiation at a greater rate than long-lived ones, and because of this tend to be more dangerous in the short-term, weight for weight. However, when samples of short- and long-lived nuclides containing different quantities of material but emitting radiation at the same rate are compared, it is the long-lived ones which are more dangerous. This is because they will continue to emit radiation at a high rate for a long time, whereas the rate of emission of radiation from the short-lived nuclides will quickly decrease. The degree of danger from a particular isotope also depends on the nature and energy of the radiation it emits.

So, in considering the effects of radioactive materials, it is necessary to look at the rate of emission of radiation at any particular time, its energy and the period (which may be thousands of years) over which the radiation will continue to be emitted at significant rates. The unit for measuring the rate of emission of radiation is the *curie,* abbreviated as Ci.

The radiation produced during radioactive decay can remove electrons from (i.e. ionise) the atoms of material exposed to it, sometimes resulting in permanent chemical changes in this material. If such changes occur in living tissue, they may affect the behaviour of its constituent cells. The biological effects of radiation on people are determined by the amount of ionisation produced by radiation absorbed in sensitive tissues.

The *rad* is the unit used to measure the energy of radiation absorbed by matter. Doses absorbed by the human body, expressed in rads, can be multiplied by a 'quality factor' which takes into account the various biological effects of different kinds of radiation. The resulting 'dose equivalent' is expressed in rems. The *rem* is the unit generally used in radiological protection standards.

Fission

When an atom decays radioactively, it emits a particle that made up only a small portion of its original nucleus. If it undergoes fission, by contrast the nucleus splits into two almost equal parts and also releases several neutrons. In addition, a very large amount of energy is released, ultimately appearing as heat and the energy of gamma radiation.

The neutrons ejected following fission are high energy or fast neutrons. Fission may sometimes occur in a nucleus of uranium-235 when it is struck by a fast neutron. However, fission occurs much more readily if the uranium-235 is struck by a neutron that has been slowed down by collisions with other nuclei. Such a slow neutron is termed a low energy or thermal neutron. If a neutron strikes a uranium-238 nucleus, on the other hand, fission is most unlikely. Instead, the uranium-238 may take in the neutron and become, via two beta emission steps, plutonium-239 ($^{239}_{94}Pu$). This isotope behaves in much the same way as uranium-235; when struck by another neutron it may undergo fission.

Naturally-occurring thorium-232 behaves in a similar manner to uranium-238. After capturing a neutron and emitting two beta particles, it becomes uranium-233, another ready candidate for fission. These two neutron-capturing isotopes are said to be 'fertile', while those which undergo fission are said to be 'fissile'.

Nuclei do not always split in the same way when fission occurs; individual atoms of one fissile substance can break up into different pairs of atoms. These leftovers, known as the fission products, are generally highly radioactive. Most are short-lived, but a few have quite long half-lives—notably strontium-90, with a half-life of 28 years and caesium-137 with one of 30 years.

The number of neutrons given off by a nucleus during fission varies with the way the nucleus splits. For uranium-235, the average is 2.5 neutrons. As this number is greater than one, a chain reaction is theoretically possible, with some of the emitted neutrons triggering further fissions. If new fissions are in fact set off by an average of more than one of the emitted neutrons, the rate of fissions, and heat output, increases exponentially. But if matters are arranged so that each fission produces on average just one new fission, the chain reaction continues at a steady rate with continuing energy release. This is what happens in a smoothly-running nuclear reactor.

Figure 2

How nuclear power stations generate electricity

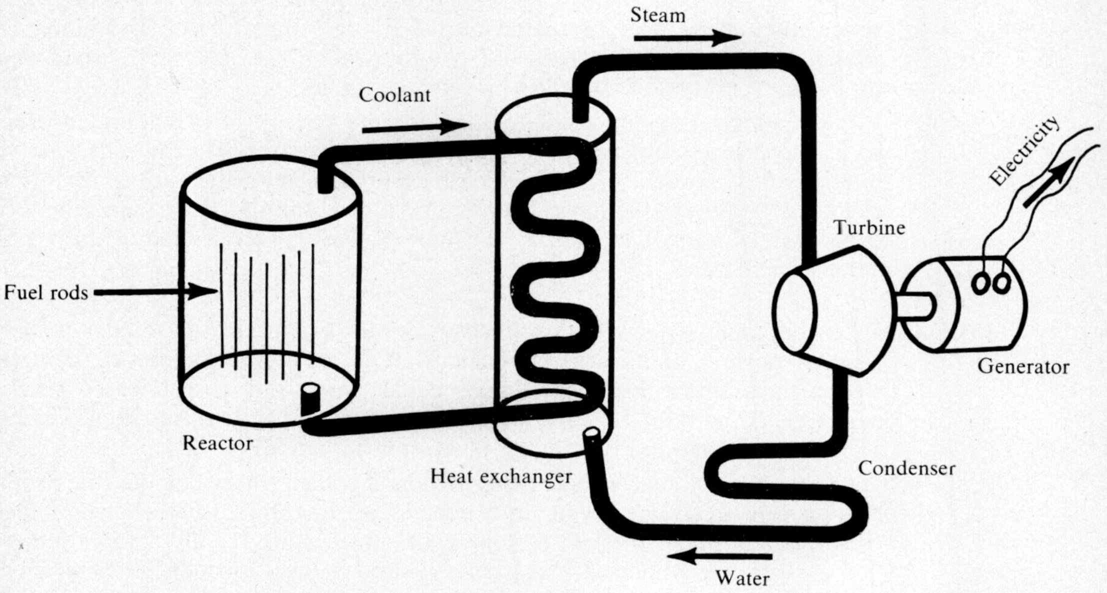

The BWR and SGHWR generate steam in the reactor, dispensing with the heat exchanger step.

Fusion Fusion is the reverse of fission in that two light atomic nuclei come together to form one heavier nucleus. However, as in fission, very large amounts of energy are given off. Fusion is the process that occurs in the sun and in hydrogen bombs. The temperatures involved are staggering—100 000 000° C or above. A controlled fusion reaction has not yet been achieved.

Nuclear reactors Many types of experimental and military fission reactors have been built to perform a variety of different functions. Power reactors, however, have one main purpose—to produce useful energy in the form of heat. All currently available commercial power reactors are designed to supply this heat in steam. Thus they have the same function as a boiler. Nuclear reactors fall into broad categories—'thermal' for those maintaining the chain reaction with slow neutrons and 'fast' for those relying on fast neutrons.

In the first category, fission is induced by slow neutrons that have lost most of their original energy by collision with nuclei of a substance known as a moderator. The slowed-down neutrons readily fission uranium-235, whilst avoiding capture by the large excess of uranium-238 in the fuel. Under these conditions, a chain reaction can proceed in natural uranium or fuel that has been slightly enriched in uranium-235 content from the 0.7 per cent occurring in natural uranium to perhaps 2-3 per cent. Commonly-used moderators are ordinary water (known as light water), heavy water (its hydrogen is the heavy $^{2}_{1}H$ isotope, deuterium), and graphite.

'Fast' reactors use uranium highly enriched in uranium-235, or high concentrations of other fissile material, as fuel. Because the concentration of fissile nuclei is so high, the chain reaction can be sustained by fast neutrons and no moderator is needed.

The region of a reactor where the heat is generated is called the core. In some reactors the core, which contains the fuel, is enclosed in a 'pressure vessel' of heavy welded steel or prestressed concrete. In others, the core materials are distributed in an array of much smaller 'pressure tubes'.

The heat produced is carried away by liquid or gaseous coolants circulating through the core. In all but two types of nuclear power station, the hot coolant passes through a heat exchanger in which water in another circuit then boils to produce steam. This steam operates a conventional turbine-generator system which produces the electricity. The two exceptions, the boiling water reactor (BWR) and the steam-generating heavy water reactor (SGHWR), dispense with the heat exchanger step. Their coolant water boils to produce steam for the turbine. (See Figure 2.)

The rate of fission in a reactor is controlled by rods of material with a large capacity to absorb neutrons. The deeper these control rods are inserted in the reactor, the more neutrons they absorb. Partial withdrawal of the rods allows enough neutrons to circulate to start the chain reaction. Full insertion stops the reaction.

Brief descriptions of reactor types follow:

Gas-cooled Reactors

Magnox Reactor

Fuel rods of natural uranium metal encased in 'magnox', a magnesium alloy, are inserted in a structure of graphite blocks which constitutes the moderator. Pressurised carbon dioxide is the coolant.

Advanced Gas-cooled Reactor (AGR)

This is a development of the Magnox reactor. The main differences are that the fuel is slightly enriched (to about 2 per cent uranium-235), and is in the form of uranium dioxide (UO_2). This allows a higher proportion of the uranium-235 to be consumed before the fuel rods are replaced and enables the reactor to operate at higher temperatures, giving greater thermal efficiency.

High Temperature Gas-cooled Reactor (HTGR)

Although gas-cooled and graphite moderated, this is a very different type of reactor. The fuel is uranium dioxide or carbide, more highly enriched in uranium-235. Thermal efficiency is increased because the coolant gas, helium, is allowed to reach much higher temperatures. A substantial quantity of fertile thorium-232 may be mixed with the uranium fuel. This captures neutrons to become uranium-233, which is fissile and makes an increasing contribution to the reactor's fuel supply.

Water-cooled Reactors

In these reactors, water—either light or heavy—is used as coolant and as moderator. Most of the world's commercial power reactors use light water and are called light water reactors (LWRs).

Boiling Water Reactor (BWR)

In concept, this is the simplest of all types of reactor. The same light water serves as coolant and moderator and provides the steam that drives the

Table 1

POWER REACTORS

CLASS	Gas cooled	Gas cooled	Gas cooled	Water cooled	Water cooled	Water cooled	Water cooled	Fast breeder
NEUTRONS	Slow (thermal)	Slow (thermal)	Slow (thermal)	Slow (thermal)	Slow (thermal)	Slow (thermal)	Slow (thermal)	Fast
MODERATOR	Graphite	Graphite	Graphite	Light water	Light water	Heavy water	Heavy water	None
COOLANT	Carbon dioxide	Carbon dioxide	Helium	Light water	Light water	Light water	Heavy water	Liquid sodium or sodium-potassium mixture
USUAL FUEL	Natural uranium metal	Slightly enriched uranium oxide	*Highly enriched uranium oxide or carbide	**Slightly enriched uranium oxide	**Slightly enriched uranium oxide	**Slightly enriched uranium oxide	Natural uranium oxide	*Uranium-plutonium oxide mixture
REACTOR TYPE	Magnox	AGR	HTGR	BWR	PWR	SGHWR	PHWR / CANDU	LMFBR

AGR = Advanced gas-cooled reactor
HTGR = High temperature gas-cooled reactor
BWR = Boiling water reactor
PWR = Pressurised water reactor
SGHWR = Steam generating heavy water reactor
PHWR = Pressurised heavy water reactor
CANDU = Canadian deuterium uranium reactor
LMFBR = Liquid metal fast breeder reactor

* Other fuel systems are possible, in particular uranium-233—thorium-232
** Mixed uranium and plutonium oxides (MOX) could also be used

turbine. The fuel is slightly enriched (typically 2.4 per cent uranium-235) uranium dioxide.

Pressurised Water Reactor (PWR)
Light water, pressurised to prevent it boiling, serves as coolant and moderator. The hot water is used to produce steam in a second circuit at lower pressure. The fuel, again uranium dioxide, is enriched to a slightly higher level—about 3 per cent uranium-235.

Pressurised Heavy Water Reactor (PHWR)
Heavy water serves as both moderator and coolant. It is pressurised to prevent boiling, like the light water moderator-coolant in the PWR, and produces steam in a second circuit in the same way. The uranium dioxide fuel is not enriched. The most common type of PHWR is the Canadian CANDU.

Steam-Generating Heavy Water Reactor (SGHWR)
The moderator is heavy water and the coolant light water. The coolant boils, as in the BWR, to provide steam directly for the turbine. The fuel is uranium dioxide enriched to about 2 per cent uranium-235.

Both PHWR and SGHWR are pressure tube systems, avoiding the necessity to use very large pressure vessels like those used in the PWR and BWR systems.

Fast Breeder Reactors (FBR)

All the reactors described so far are 'thermal'. 'Fast' reactors have some theoretical technical advantages over them, and some disadvantages. So far they have not been developed beyond the prototype stage. A large proportion of the neutrons liberated at each fission is available for converting fertile into fissile material in fast reactors because no moderator is used, and the reactor core is very compact. The fuel for existing fast breeder reactors is a mixture of about 20 per cent plutonium oxide with natural or depleted uranium, also as the oxide. The name 'breeder' derives from the fact that they are designed to produce more fissile material, usable as reactor fuel, than they consume. But they are not perpetual motion machines. What happens is that neutrons released by the reactor fuel convert fertile material, for example uranium-238, located mainly in a blanket around the reactor core, to fissile material (plutonium-239 if the blanket is uranium-238) faster than the fissile material in the fuel is converted to non-fissile material. Fast breeders should enable very much more energy to be produced from uranium than is possible with thermal reactors.

Liquid Metal Fast Breeder Reactor (LMFBR)
Liquid sodium is used as the coolant in the prototypes now operating. This liquid metal can cope with the very large amounts of heat produced in small volumes in the reactors. The primary fuel in existing prototypes is a mixture of plutonium isotopes; uranium-235 or uranium-233 could also be used. Fertile material—uranium-238 or thorium-232 is included for breeding new fissile material. The coolant absorbs neutrons and becomes intensely radioactive when it passes through the reactor. So, like the reactor, the cooling circuit has to be kept inside thick shielding. A heat exchanger within this shielding transfers heat from the coolant to another liquid metal circuit, which in turn transfers its heat to a water circuit in a second heat exchanger outside the shielding. The water boils, and the generated steam drives the turbine.

Gas-Cooled Fast Breeder Reactor (GCFBR)

This is a possible future development, overcoming the problems posed by the propensity of the liquid metal coolant to capture neutrons. The coolant envisaged is helium, circulating at high pressure.

Fusion Reactors These are another possible development, probably further in the future. The potential contribution of fusion reactors, if the very large problems involved in their development are overcome, is discussed in Chapter 6.

The uranium fuel cycle From mining to the final disposal of waste materials, uranium undergoes a complex sequence of processes collectively termed the fuel cycle. Different stages of the cycle may take place hundreds or even thousands of miles apart. After emerging from the mine, the ore is crushed, and then chemically treated to produce yellowcake, which is more than 90 per cent U_3O_8. For reactors requiring enriched uranium, this is then converted to gaseous uranium hexafluoride, UF_6, before having its uranium-235 content boosted by enrichment. In the next step, fuel is fabricated to suit the requirements of reactors. Then it undergoes fission in a reactor, releasing some of its nuclear energy. Finally the used fuel elements are reprocessed, separating materials such as uranium and plutonium that can go back into the cycle from wastes that have to be disposed of. Neither the return of uranium and plutonium to the fuel cycle nor the final disposal of these highly radioactive wastes has begun on a commercial scale, so the 'nuclear fuel cycle' is not yet, strictly speaking, a cycle. (See Figure 3.)

Chapter 10 deals with hazards that arise at different stages in the fuel cycle. This section briefly describes the processes involved.

Mining The ore is mined using techniques identical with those used in mining other ores. Open-cut or underground methods are selected depending on, among other things, the depth of the ore body and the amount of material overlying it.

Production of yellowcake (milling) After being crushed and ground, the ore is chemically treated to extract the uranium as yellowcake, consisting of more than 90 per cent U_3O_8. This is then crushed to form a fine powder. The waste material, called 'tailings', is normally deposited behind a dam, for indefinite storage.

Conversion Enrichment plants require the uranium they treat to be in the form of a gas, and the compound invariably used is uranium hexafluoride, UF_6. Two steps are involved in producing it—purification of the uranium and chemical conversion to the fluoride compound. They are usually carried out as stages in a single overall process known as conversion.

Enrichment Boosting the proportion of fissile uranium-235 in a sample of natural uranium is a very difficult operation. What makes it possible is the slight difference in mass between the fissile isotope and the dominant uranium-238. This difference leads to an even smaller difference in the average speeds of molecules of UF_6 which, like those of any gas, are always moving round in a random manner. The average speed of a UF_6 molecule containing the lighter uranium isotope is about 0.4 per cent faster than that of a molecule containing uranium-238.

The established enrichment process is gaseous diffusion. The uranium hexafluoride gas is forced by pumping to diffuse through porous membranes, and the faster-moving molecules move through slightly faster than the slower ones. Gas that goes through the membranes is slightly enriched in the lighter isotope, and is separated from the correspondingly impoverished gas remaining behind. The depleted product is known as the 'tails'.

To reach the 2-3 per cent uranium-235 enrichment level required for most types of reactor, the uranium hexafluoride goes through some thousands of stages of diffusion, in a cascade arrangement. Each stage is a metal-walled cell, divided by a thin membrane of porous metal. As enrichment proceeds, the total mass of uranium hexafluoride passing through successive stages decreases, so the cells are made successively smaller. As a consequence, as much pumping capacity is said to be required for enrichment from the natural 0.7 per cent to 4 per cent as for further enrichment from 4 per cent to nearly 100 per cent uranium-235. Gaseous diffusion plants use very large amounts of electrical power.

An alternative enrichment technology now under active development, the gas centrifuge method, is claimed to have a much smaller consumption of electricity for a given amount of enrichment. When uranium hexafluoride enters a spinning centrifuge, the heavier molecules tend to drift to the outside leaving the lighter molecules containing uranium-235 nearer the centre. The gas is divided into two streams, one slightly enriched in the lighter molecules containing uranium-235 and the other slightly depleted. The degree of separation that can be achieved in one centrifuge is small, so the uranium hexafluoride is gradually enriched as it passes through a succession of centrifuges.

Another method under development is the separation nozzle process. In this process, jets of uranium hexafluoride, mixed with hydrogen, are pumped through curved tubes. The lighter uranium hexafluoride molecules tend to move off at slightly greater angles than the heavier ones, and the enriched and impoverished streams are separated at the nozzle. As in the other processes, the gas has to pass through many stages before it reaches required levels of enrichment. The basic mechanism of separation is the same as that in the centrifuge process. But a major advantage of this process over centrifuge enrichment is said to be the absence of the mechanical problems associated with highly-stressed rotating machines. A drawback is the power consumption which at the present state of development is said to be slightly greater than that of existing U.S. diffusion plants.

Lasers, which produce beams of intense coherent radiation, are the basis of another enrichment process now at an earlier stage of development. Evidently lasers can be tuned so finely that their radiation ionises uranium-235 hexafluoride molecules while not ionising uranium-238 hexafluoride. The electrical charge given to the molecules containing uranium-235 would be used to separate these out. If the process turns out to be technically possible, a single stage of laser enrichment may be able to separate almost completely the uranium isotopes. This would offer a short cut to weapons material.

Fuel fabrication

The fuel used in most reactors, whether enriched or not, is uranium dioxide. So enriched uranium hexafluoride, or yellowcake if natural uranium is the fuel, has to be converted into uranium dioxide powder. This is then compressed into small, cylindrical pellets. Columns of these pellets, up to several

Figure 3

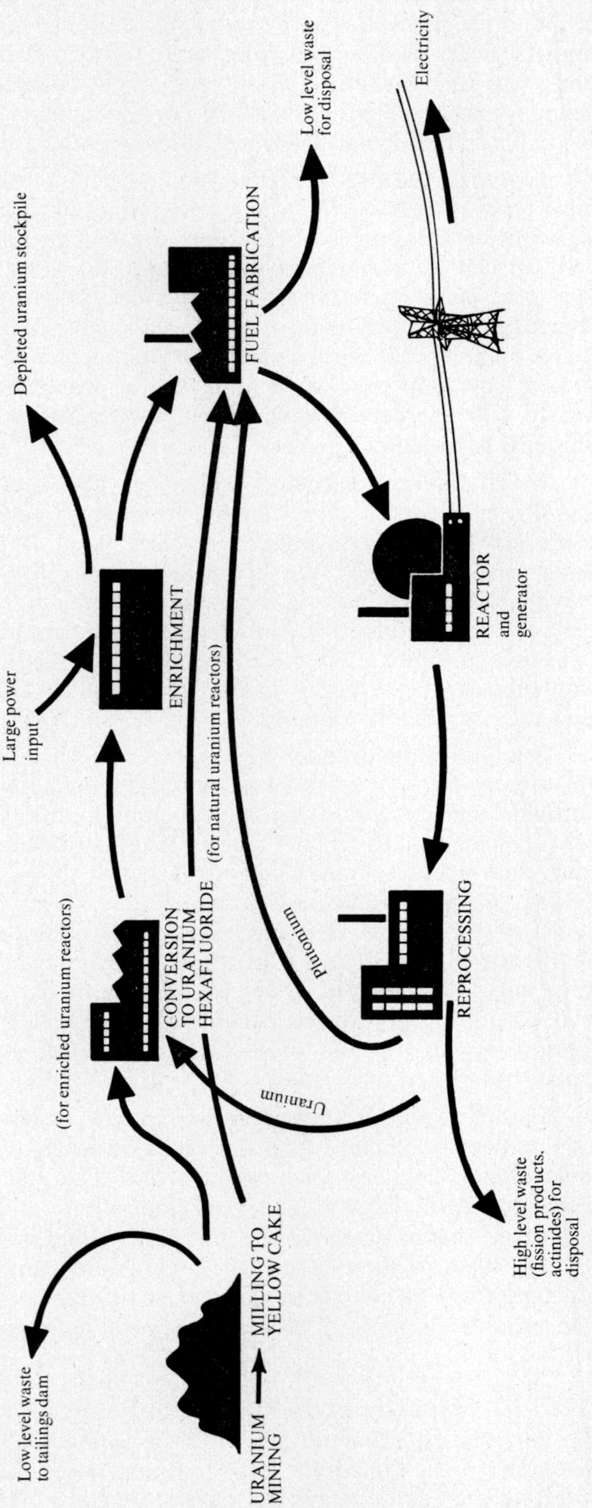

The nuclear fuel cycle

metres long, are inserted into thin-walled tubes known as fuel pins or rods. These tubes have to be able to withstand high temperatures, and are usually made of stainless steel or a zirconium alloy.

The fuel has to be fabricated under conditions of extreme cleanliness, to keep out impurities that could absorb neutrons needed to keep fission going. Fuel containing plutonium poses special problems in fabrication, as it is more toxic than uranium and greater care is needed to ensure that enough does not accumulate in one place to cause an unwanted chain reaction.

Reactor operation

The fresh fuel rods are assembled in the reactor in a carefully engineered structure, known as the reactor core, where they are held in precise geometrical relation to the control rods, the moderator and the gaps for the passage of coolant fluid. The fuel rods are left in position for up to three years—the precise time depending on the type of reactor and many other detailed operational characteristics—before being removed and replaced by fresh rods, an operation known as refuelling.

Nuclear reactors may be divided into two broad categories, according to whether they have to be shut down for refuelling. Both PWRs and BWRs have to be shut down. Others, including Magnox and CANDU reactors, can be refuelled while 'on-load' (operating), and their fuel rods are usually replaced continually.

Fuel reprocessing

Spent fuel elements removed from a reactor contain unused fuel, a wide range of fission products, fissile and non-fissile material produced by neutron capture, and radioactive decay products. They are extremely radioactive, much more so than the material encountered at any other stage of the fuel cycle.

The products of neutron capture and radioactive decay include neptunium-237, plutonium-238, -239, -240 -241 and -242, americium-241 and curium-242 and -244. These, and some other isotopes present in smaller quantities, are actinides—the name given to actinium, whose atomic number is 89, and to all elements with atomic numbers above 89. Some of the actinides are long-lived, notably plutonium-239, -240 and -242 which have half-lives of 24 400, 6 760 and 379 000 years respectively, and americium-241 and -243 whose half-lives are 458 and 7 650 years respectively.

Most of the fission products are short-lived, but not all. Strontium-90 and caesium-137 have already been mentioned as notable exceptions.

In reprocessing, constituents of the spent fuel elements are separated chemically. Before it begins, the fuel elements spend a few months 'cooling' under water, during which time the short-lived isotopes decay to low levels of radioactivity. Because of the intense radioactivity, reprocessing is done entirely by remote control from behind heavy shielding. Nearly all the uranium and plutonium is extracted. The other actinides and the fission products are wastes requiring disposal.

Fuel recycling

The extracted uranium is depleted in uranium-235 compared with fresh fuel, but, in the case of reactors using enriched fuel, has a higher concentration of uranium-235 than natural uranium. It has potential value as fuel, and could be recycled through reactors after conversion to uranium hexafluoride, enrichment, and fabrication into fresh fuel.

The plutonium can be used in fuel for some types of reactors instead of enriched uranium. It would be recycled directly to a fuel fabrication plant. Neither of these recycling steps is in commercial operation yet. Recycling of bred fissile material (plutonium in current systems) will be essential for the economic operation of fast breeder reactors.

Waste disposal

The radioactive wastes produced in the nuclear fuel cycle—solid, liquid and gaseous—are treated as low, intermediate or high level according to their degree of activity. Low level wastes are discharged to the environment; the other wastes are concentrated and stored in special containers. Possible hazards associated with the various wastes are discussed in Chapter 10.

4 THE PRESENT STATUS OF NUCLEAR POWER

The first experimental nuclear power reactors started producing electricity in 1954 in the U.S.A. and the U.S.S.R. Britain's Calder Hall reactors, which started up in 1956, were the first to supply electricity to a national grid. By 31 December 1975, electricity was being supplied from 157 power reactors of capacity larger than 30 megawatts electrical (MWe). Total capacity was 72 gigawatts electrical (GWe), more than three times the figure for 1970 (see Figure 5). (One GWe equals 1000 MWe or 1000 million watts electrical.)

Most of those 157 reactors are operated commercially by electric utilities, but the total includes several prototype reactors, including two prototype fast breeder reactors, which are operated by research and development agencies. In addition, a number of prototype and experimental power reactors with capacities of less than 30 megawatts electrical were operating.

Table 2 shows the location of power reactors in different regions of the world. It can be seen that at present the U.S.A. has about a third of the reactors and just over half the total capacity. All but one of the U.S. power reactors are light water reactors (LWRs), rather more than half of these being pressurised water reactors (PWRs) and the rest boiling water reactors (BWRs). For the world as a whole, light water reactors are the most important type, accounting for over 80 per cent of total capacity.

Most of those 157 reactors are operated commercially by electric utilities, which has no other type at present in commercial use. Most heavy water reactors (HWRs) are the CANDU (Canadian Deuterium Uranium) type built in Canada; several of these have been sold to developing countries. Other countries with large numbers of nuclear power stations are Japan, France, the Federal Republic of Germany, and the U.S.S.R. The Soviet Union and the other countries with centrally planned economies had about 10 per cent of total world capacity in 1975.

Altogether, nineteen countries had operating nuclear power stations at 31 December 1975; ten of these had three or fewer. The countries with nuclear power stations were Argentina, Belgium, Bulgaria, Canada, Czechoslovakia, France, the Federal Republic of Germany, the German Democratic Republic, India, Italy, Japan, the Netherlands, Pakistan, Spain, Sweden, Switzerland, the United Kingdom, the United States and the Union of Soviet Socialist Republics.

Another seven countries—Austria, Brazil, Finland, Hungary, Mexico, the Republic of Korea and Taiwan—had power reactors under construction. Six more—Iran, Luxembourg, the Philippines, Poland, Romania and Yugoslavia—had them on order.

Prototype liquid metal fast breeder reactors (LMFBRs) have been generating electricity for several years in the U.S.S.R. and France. The Soviet prototype has a capacity of 350 megawatts electrical and the French one, named Phenix, a capacity of 233 MWe. The 250 MWe British Prototype Fast Reactor (PFR) started full power operation during 1976. Construction of a 600 MWe LMFBR is well under way in the U.S.S.R. France plans to move directly to the 1200 MWe Super-Phenix, which is regarded

24647/76—3

Figure 4

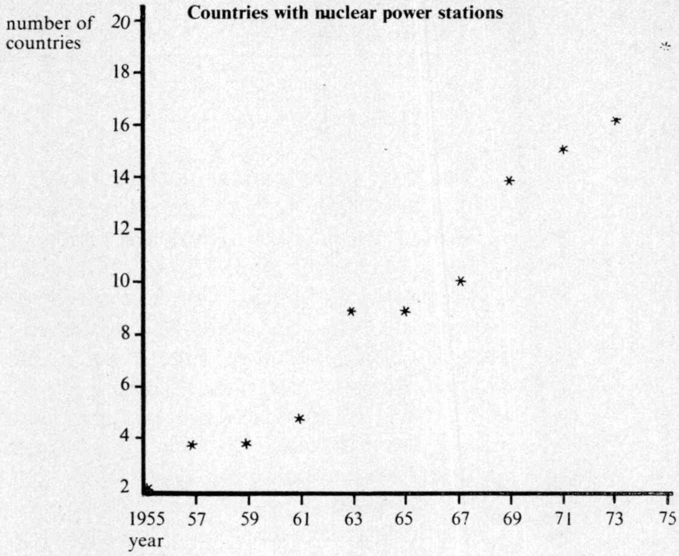

Figure 5

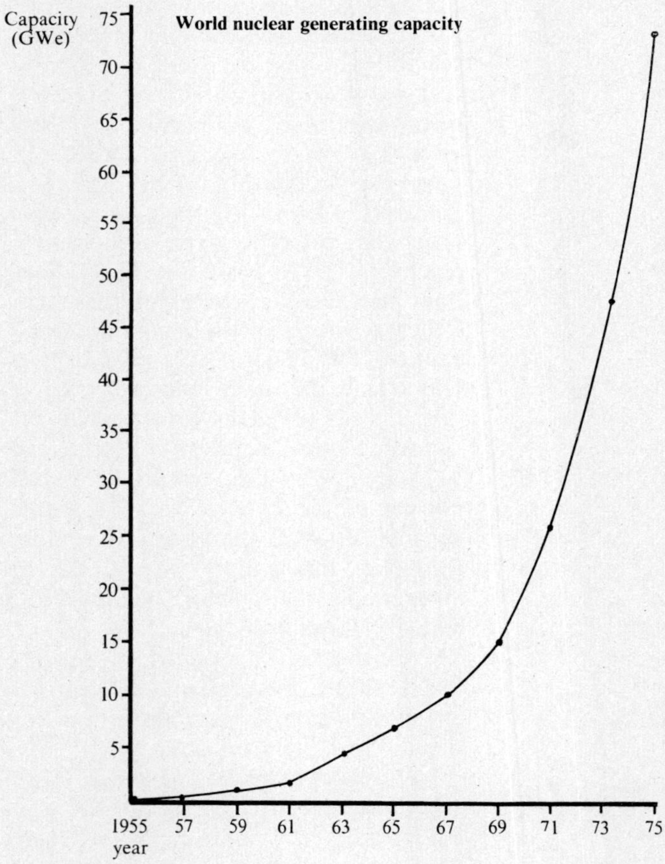

The figures for 1955 to 1973 are from IAEA.
The 1975 figures are from *Nuclear News*, February 1976.

as a commercial prototype, but construction is not expected to start for several years. Other countries with a major interest in LMFBR development are the Federal Republic of Germany, where construction of a 282 MWe prototype is well under way, and Japan, which plans to start building a 300 MWe unit next year. In the U.S.A. the proposed 350 MWe Clinch River Breeder Reactor is awaiting government authorisation for construction work to begin. All these countries, plus Italy, have small experimental fast reactors.

Small research reactors of various types are widely distributed around the world. More than 350 exist in some fifty countries. The U.S.A. has about 120 research reactors, and the Federal Republic of Germany, the U.S.S.R., the U.K., France and Japan between twenty and thirty each. Almost all the other countries have five or fewer.

Research reactors, like power reactors, are structures within which controlled fission chain reactions take place. However, most of them have a very small thermal output of a few kilowatts or less, and they do not generally produce electrical energy. They are used for many types of nuclear physics research. Some large research reactors have thermal outputs of tens of megawatts, and may produce significant quantities of plutonium.

Australia has two research reactors at the Australian Atomic Energy Commission Research Establishment at Lucas Heights, Sydney. One of these, HIFAR, has a capacity of 11 megawatts thermal, while the capacity of the much smaller MOATA reactor is 10 kilowatts thermal. HIFAR is a heavy water reactor using high-enriched uranium fuel imported from the U.K. Its uses include the production of various isotopes for medical, industrial and other purposes.

Other components of the nuclear fuel cycle are less widely distributed than reactors. Uranium mining and milling operations are discussed in Chapter 8. Outside the countries with centrally planned economies, there are only five commercial plants for the conversion of yellowcake to almost pure uranium hexafluoride—two in the U.S.A., and one each in the U.K., France and Canada.

There are three enrichment plants in the U.S.A. and one each in the U.K. and France, all built originally as part of those countries' military programs and using gaseous diffusion technology. The British and French plants are small, and until a few years ago the plants in the U.S.A., owned by the Government through the U.S. Atomic Energy Commission (now the Energy Research and Development Administration) had a virtual monopoly on the provision of enrichment. However, the U.S.S.R. is now providing an alternative source and has won a number of contracts to enrich fuel for power stations in Western Europe. China also has a military enrichment plant, probably small.

If nuclear power capacity grows according to the latest projections available to the Commission, present enrichment capacity will be insufficient to meet expected fuel requirements within a few years. For this reason an international consortium, known as Eurodif (France, Italy, Spain, Belgium and Iran) is building a large, commercial gas diffusion enrichment plant in France. Another consortium, known as Urenco (U.K., Federal Republic of Germany and the Netherlands) is operating pilot-scale gas centrifuge plants in the U.K. and the Netherlands, and plans to build a commercial plant in each country. Japan and Australia have gas centrifuge research and development programs.

Table 2

Nuclear power reactors around the world at 31 December 1975
(units smaller than 30 MWe are omitted)

	Operating			Being Built		On Order		Total		
	Number	*Capacity (GWe)*	*Percentage*	*Number*	*Capacity (GWe)*	*Number*	*Capacity (GWe)*	*Number*	*Capacity (GWe)*	*Percentage*
United States of America	53	36.4	50.6	62	62.8	99	111.1	214	210.3	55.6
Canada	6	2.5	3.4	6	4.1	8	5.2	20	11.8	3.1
Japan	11	5.9	8.2	12	9.3	1	0.3	24	15.5	4.1
Federal Republic of Germany	7	3.3	4.6	11	10.2	8	9.8	26	23.3	6.1
France	10	2.8	3.9	16	14.7	6	5.5	32	23.0	6.1
United Kingdom	28	5.3	7.4	11	6.5	..	..	39	11.8	3.1
Other EEC countries	9	3.6	5.0	..	..	9	9.0	18	12.6	3.4
Other western European countries	10	4.5	6.2	19	15.3	10	7.2	37	27.0	7.3
Developing countries	5	1.0	1.4	12	6.8	14	11.5	31	19.3	5.0
U.S.S.R.	12	4.6	6.4	8	5.9	5	3.9	25	14.4	3.8
Other countries with centrally planned economies	6	2.0	2.9	9	4.0	7	3.0	22	9.0	2.4
World Total	157	71.9	100.0	166	139.6	165	166.5	488	378.0	100.0

Source: *Nuclear News*, World List of Nuclear Power Plants, February 1976.
Note: A more complete version of this table can be found in Appendix **B**.

The Federal Republic of Germany has developed another enrichment technology, jet nozzle, which it is planning to export to Brazil. Reports of South African developments in enrichment suggest that it may be building a plant based on a related technology. All these projects are being financed by the governments involved. However, the U.S. Government has proposed that new enrichment capacity in that country be financed by private companies. Discussions between a number of companies, potential customers and the Government are proceeding, but no firm plans have been announced.

Fuel fabrication is technically a less complex operation than most other stages of the fuel cycle. At present there are over 40 plants in 14 countries, excluding the countries with centrally planned economies; in fact almost every country with a power reactor also has a fuel fabrication plant.

The metallic fuel used in many gas cooled reactors is reprocessed at plants in the U.K. and France. There are no plants for reprocessing oxide fuel (used in all other types of commercial reactor) currently in commercial operation. Because little reprocessing is being done, the number of spent fuel elements stored in cooling ponds is growing rapidly.

Oxide reprocessing plants in the U.S.A. and the U.K. have been shut down for modifications and are not expected to re-open for several years. Another plant in the U.S.A. and one in France are still being built and also will not begin operating for several years. A smaller one in Japan is close to completion. The Federal Republic of Germany plans to build a large production-scale plant. A third U.S. plant ran into severe technical difficulties while still under construction and has been abandoned for the time being.

However, all these countries and others, including India, Italy, Spain, Yugoslavia, Argentina and Taiwan, are operating or building experimental or pilot scale reprocessing plants. The Federal Republic of Germany plans to sell a pilot plant to Brazil, and France has recently announced a contract to sell a small reprocessing plant to Pakistan. The reprocessing situations in the U.S.S.R. and other countries with centrally planned economies are not known.

Plutonium recovered during reprocessing of power reactor fuel is at present being used as fuel in experimental and prototype fast reactors. Commercial recycling of plutonium and uranium from reprocessing has not begun, and the economic viability of reprocessing has not been demonstrated. If it does begin, the plutonium, as well as the uranium, could be used as fuel in thermal reactors.

Because very little fuel is being reprocessed at the present time, little high level waste is being produced. High level waste produced in previous years in the U.S.A., the U.K. and France is stored as a liquid in special tanks, as are the much larger quantities of high level waste from military plutonium production reactors. A number of options for disposing of high level waste are now being considered; the various options are discussed in Chapter 10. Authorities in the U.S.A. do not expect to make a decision on ultimate disposal for more than a decade.

Under the Treaty on the Non-Proliferation of Nuclear Weapons (NPT), arrangements have been made to try to prevent nuclear material from being diverted for use in the development and manufacture of nuclear weapons. Similar 'safeguards' arrangements have been developed independently of the

NPT by the International Atomic Energy Agency (IAEA), which also operates the NPT safeguards system (see Chapters 12 and 13).

Under the NPT arrangements, all fuel cycle facilities in non-nuclear weapon countries, except those concerned only with source material, should be covered by safeguards. Uranium mines and mills are not covered. Two countries with nuclear weapons, the U.S.A. and U.K., have offered to place their civil nuclear fuel cycle facilities under NPT safeguards. The U.K. offer was accepted recently by the IAEA Board of Governors. An agreement giving effect to the U.S. offer is to be considered by the Board in the near future. Facilities in the other nuclear-weapon countries—the U.S.S.R., France and China—are not safeguarded, and it is believed that there is no proposal that they will be.

A complete list of facilities under and not under IAEA or NPT safeguards could not be obtained by the Commission. The evidence indicates that most fuel cycle facilities in countries without nuclear weapons are safeguarded. However, the following are among facilities which are not under such safeguards:

- In India, the 40 megawatts thermal Cirus research reactor, supplied by Canada but using indigenous uranium fuel, which produced the plutonium used in the nuclear device exploded by India in 1974; plus three other small research reactors, a fuel fabrication plant and a reprocessing plant.
- In Israel, the 26 MWth Dimona research reactor supplied by and using fuel from France.
- In Egypt, the 2 MWth Inshas research reactor supplied by and using fuel from the U.S.S.R.
- In South Africa, a prototype uranium enrichment plant.
- In Spain, the 480 MWe Vandellos gas-cooled power reactor, jointly operated and controlled with France.

5 WORLD ENERGY CONSUMPTION

One of the factors that will influence future demand for uranium for electricity generation is the rate of growth of energy use, and particularly of electricity use, around the world. The evidence before the Commission indicates that the use of energy will continue to grow during the remainder of this century, but generally at slower rates than have prevailed in most of the post-war period.

Likely influences on future energy growth rates include population growth rates, levels of economic activity, the relative costs of energy, the success of energy conservation programs, and concern about various social and environmental effects of continually expanding energy use. In this chapter, we relate predictions of energy growth rates to past trends, and discuss evidence presented to the Commission on energy conservation and on the possible wider consequences of continued energy growth.

Historical growth in use of energy

An evaluation of past trends in energy consumption is essential to an understanding of the contribution which uranium may make in the future as a fuel for electricity generation. In all countries for which data are available, both liquid fuels and electricity have experienced high rates of growth in production and consumption during periods of sustained economic growth, particularly between the Second World War and the early 1970s. Growth in the use of solid fuels, particularly coal, has taken place at slower rates, and the use of coal has declined in a number of developed countries since the Second World War.

The Commission heard a number of submissions dealing with the relation between levels of economic activity and growth, and the use of energy, particularly electricity. The data suggest that growth of total energy use has been slower than economic growth in the older economically developed economies, such as the U.S.A. and Western Europe, and faster in countries which reached an economically developed stage more recently, such as Japan and Australia.

In the developing countries taken as a whole, the energy growth rate has been substantially greater than the economic growth rate, as measured in terms of statistical indicators such as gross national product. However, the comparatively low levels of energy consumption in these countries need to be taken into account in interpreting these growth rates. Economically developed countries, excluding the centrally planned economies, are estimated to have consumed about 59 per cent of world energy production in 1975, while the developing countries with market economies, which accounted for nearly half the world's population, consumed only 10 per cent of the total. The remaining 30 per cent was consumed by countries with centrally planned economies. Since no evidence before the Commission indicates that energy developments in the centrally planned countries will have any significant influence on the demand for Australian uranium in the foreseeable future, energy requirements in those countries are not discussed in this part of the report.

The use of electricity, as distinct from energy as a whole, increased at must faster rates than economic growth between World War II and the late 1960s. Data for more recent periods show some decline in energy and electricity growth rates in the early 1970s, before the main impact was felt of the substantial oil price increases imposed by the Organization of Petroleum Exporting Countries (OPEC).

The decline in economic activity in 1974, due partly to the quadrupling of oil prices by the OPEC countries but probably due more to the recession which was already under way and anti-inflationary measures, was accompanied by a decline in overall energy use. Energy consumption in the 24 industrialised countries which are members of the Organization for Economic Co-operation and Development (OECD) fell by an average of about 2 per cent in 1974, while the level of economic activity declined marginally. A further fall of about 3 per cent in energy use occurred in these countries in 1975, accompanying a decline of about 2 per cent in the overall level of economic activity.

Economic recession, higher energy prices and attempts at energy conservation all influenced energy use in 1974 and 1975. Recent reappraisals of energy projections and policies in the economically advanced countries appear to take these factors into account, at least to some extent. However, it seems reasonable to expect further analysis of these factors to be followed by more changes in energy projections and policies; there is inevitably a considerable time lag between events and the development of policies and forecasts influenced by them.

The Commission recognises that there is scope for considerable difference of opinion about likely rates of growth of economic activity and energy consumption in the future. Current doubts about trends in economic activity and in relative prices for different forms of energy make it extremely difficult to estimate future energy consumption patterns. In any case, it needs to be borne in mind that the projections which are published do not represent 'plans' in any formal sense; they are only predictions of likely trends in energy use. The economies of the countries concerned are principally market economies in which demand, supply and relative prices play a major part in determining the course of events.

Projections of energy and electricity usage in developed countries

Projections by the OECD Secretariat in *Energy Prospects to 1985*, published in 1974, serve as a guide to possible levels of energy use in developed countries (excluding the centrally planned economies) in the next decade. The Secretariat sought to establish the extent to which higher oil prices would affect energy consumption in OECD member countries, and assumed that average rates of growth of economic activity in the next decade would be somewhat lower than the average experienced in the post-war period. The study projected a slower rate of growth of total energy use compared with economic activity using pre-1973 oil price levels, and expected the oil price rise to produce a further slowing in the energy growth rate. Some diversion from the direct use of petroleum to the consumption of electricity was expected to occur.

Estimates of 1985 levels of electricity use were based on assumptions of slower rates of growth in economic activity, and on anticipated costs of electricity compared with those of other forms of energy. An expected growth in nuclear power generation resulted from the economic advantage that nuclear power stations were anticipated to have over fossil-fuelled plants, particularly over the oil-fired plants which make up a large proportion of the additional generating capacity installed in Western Europe and Japan since World War II.

More recent and more detailed studies of the potential growth of electricity consumption in the United States and Japan are available. The U.S. Federal Energy Administration's *1976 National Energy Outlook* predicts an annual growth of five to six per cent in U.S. electricity consumption up to 1985, with a possible reduction to about 4 per cent if strong energy conservation measures are adopted. The latest published information on the anticipated growth of demand for electricity in Japan suggests an average increase of 5.6 per cent per year between 1973 and 1985.

The recent revisions of projected growth in electricity use in the United States take some account of the economic conditions which applied in 1974 and 1975, including the deferments of orders, particularly for nuclear power stations, during the economic recession. Similar revisions relating to Western European and Japanese projections were not available to the Commission. A good deal of uncertainty therefore remains about the levels of electricity consumption which will be reached by 1985.

The projections referred to of energy and electricity growth in the next decade, published by national and international bodies, are based on a belief that persistent growth in economic activity will be resumed in the countries concerned, and that this growth will be accompanied by increased use of energy. Although average growth rates in energy use are expected to be below the averages recorded in the post-war period, it is also generally assumed that the proportion of energy used as electricity will continue to increase. The major reason for this expectation is that electricity can be further substituted for other forms of energy, particularly petroleum, which seem likely to be subject to increasing supply constraints, making them relatively more expensive than electricity.

The Commission heard evidence that it would not be necessary for the use of electricity to grow at the projected rates. It was suggested that acceptable levels of economic activity could be reached without the predicted growth in electricity consumption. The Commission's attention was drawn to a study carried out as part of the Ford Foundation Energy Policy Project, *A Time to Choose,* published in 1974. As one part of this study, a model of the United States economy was used to simulate alternative developments in the period from 1975 to 2000. The results obtained from this model suggest that high energy growth rates are not necessarily required for high levels of economic activity and employment. One of the study's principal conclusions—that the relationship between measured economic growth and energy growth is subject to considerable variation under alternative policies —was generally endorsed by opponents of uranium mining in their submissions to the Commission.

In a number of cases, witnesses went further and suggested that growth of economic activity, as measured by growth of gross national product, is not necessarily a goal which societies will continue to pursue. The Commission notes that there is evidence that societies may come to value more highly in future things not included in conventional measurements of economic activity, and that many people hold the view that zero economic growth, as conventionally measured, is both possible and consistent with other goals of society. It also notes that the Energy Policy Project argues that sustained growth in economic activity in the U.S. could continue even if energy consumption increased at much lower rates than in the past.

It is obvious that attempts to forecast the course of events over the rest of the century are subject to wide margins of error. However, it seems clear that use of energy and electricity will be affected not only by rates of growth of economic activity and other economic considerations, but also by a probable decline in the rate of growth of population. The rate of growth of energy use will also be influenced by the extent to which policies aimed at energy conservation are successful.

Energy conservation

Energy conservation was assigned a very important role, along with renewable sources of energy, by proponents of non-nuclear energy in their submissions to the Inquiry. No witness disputed that conservation, aimed at achieving social objectives with the least possible energy use, is desirable, but there was disagreement about the size of its possible impact. Many of the arguments revolve around whether particular conservation measures will be implemented rather than whether they could be.

In the United States, a number of studies have shown considerable potential for energy conservation. For example, a comparative study of energy consumption in the U.S.A. and the Federal Republic of Germany, commissioned by the U.S. Federal Energy Administration, found that energy use per unit of national income in the Federal Republic of Germany is only 66 per cent of that in the U.S.A. despite the similarity of real income per head in the two countries. The revised version of ERDA's *A National Plan for Energy Research, Development and Demonstration,* published in 1976, gives what it calls 'significantly increased' priority to conservation technology in the period to 1985. Among the reasons given for this change are that it typically costs less to save a barrel of oil than to produce one through the development of new technology, that energy conservation has a lower environmental impact than energy production and use, and that conservation technologies can be implemented comparatively quickly.

The Commission notes that in 1975 the Government of Sweden set an energy growth target of 2 per cent per annum up to 1985, falling to zero by 1990. Sweden's official energy strategy gives high priority to conservation technologies. The Commission was told that the Swedish Government planned to institute an energy tax, provide loans and grants for energy-saving modifications to houses and factories, amend building legislation to discourage energy consumption, and institute a scheme of municipal energy planning incorporating conservation. Other evidence demonstrated that large reductions in the energy used in domestic and commercial buildings could be achieved by relatively small changes in architectural practice.

The evidence shows that the major energy consuming nations have embarked on energy conservation programs of varying intensity, and that these are being given high priority. The Commission notes that the long-term co-operation program adopted by OECD's International Energy Agency in 1976 has, as one of its principal aims, the promotion of energy conservation, and provides for co-ordination of national efforts and the establishment of co-operative efforts in this field. While it is difficult to predict the likely effect on the level of overall energy consumption which conservation programs will achieve in the next decade, it seems certain that they will reinforce past and likely future increases in relative energy prices in restraining growth in the use of energy. Energy conservation therefore seems likely to have a significant effect on total energy consumption by the end of the century.

Wider consequences of continued energy growth

Although a number of countries may experience lower average energy growth rates in the latter part of this century than in the past, the evidence suggests that a great deal of thought will have to be given to the cumulative economic, social and other environmental aspects of continued growth in electricity consumption in economically advanced countries. Such growth inevitably involves substantial increases in the rate of building and commissioning of generating, transmission and distribution facilities.

It was pointed out in evidence that nuclear energy is much less flexible, at least with presently available and foreseeable future technology, than coal, oil and natural gas in that it can be used economically only to generate electricity, and only in large power stations. It was argued that, because of this, an increasing proportion of total energy consumption is likely to be supplied as electricity in the future if nuclear power programs grow as projected.

Some critics of nuclear power drew attention to the possible social consequences of a society where electricity distributed through a centralised grid is the major source of energy. It was said that this would require administration by a remote and bureaucratic technical elite, lead to a great concentration of political and economic power, and be vulnerable to large and expensive technical mistakes and failures. It was also submitted that the large scale and complexity of nuclear power will reduce the opportunity for greater public control of decision-making and may threaten democratic procedures and civil liberties.

It was suggested that these social consequences will be of such significance that they constitute an argument for rejecting nuclear power. Although no contrary evidence was presented to it, the Commission is aware that conflicting views exist about whether such consequences will occur and whether they will be more severe if nuclear power becomes a more important source of electricity. It should be noted that the effects concerned would be additive to those, including the restriction of civil liberties, which might result from measures to control diversion of nuclear material and proliferation of nuclear weapons; such effects are considered in Chapters 13 and 14.

On balance, the evidence suggests that growing recognition of the overall social and economic implications of continuing growth in energy use is likely, and therefore many countries may be forced to take account of them in their energy policies. It is conceivable that more of the costs associated with these consequences will be reflected in the costs of energy supplied to consumers in future, in keeping with the 'polluter pays' principle. It appears that additional policy measures may be necessary to achieve desired reductions in the growth of energy consumption.

Possible effects of continued energy growth on climate

The end result of what is generally described as consumption of energy is not the actual destruction or loss of the energy, but its conversion to a form which is no longer generally useful—heat at a low temperature. This joins radiation from the sun in heating the earth. Unlike the use of fossil and nuclear fuels, the use of solar energy does not significantly add to the amount of heat on the earth. This is because the solar radiation is dissipated on earth whether or not some of it is used as an energy source.

Climate is primarily determined by the input of heat energy from the sun. Extra heat from non-renewable energy consumption could produce changes, for example, in average temperature or rainfall. Consumption of the fossil fuels—coal, oil and natural gas—may also influence the climate by increasing the amount of carbon dioxide and dust in the atmosphere.

Not enough is known about the atmosphere for confident predictions to be made of the actual consequences of any increase in carbon dioxide, dust or low temperature heat release. However, the evidence indicates that carbon dioxide and dust are the more important factors. Their effect, which is a direct environmental consequence of fossil fuel combustion, should be debited against those energy sources (see Chapter 11).

Low temperature heat release, to which nuclear energy contributes (see Chapter 10), is directly related to total human energy use. At present it contributes to the biosphere only about a ten thousandth of the energy input from the sun, and the global effects are generally accepted to be of little significance. But local effects over large cities can be quite pronounced. It would certainly be prudent to envisage a levelling off in energy use well before such effects become widespread.

Conclusions

Historical rates of growth in energy and electricity consumption appear to have been significantly affected by the substantial increase in petroleum and other energy prices in the last few years. Although there is considerable doubt about the extent to which depressed economic conditions have influenced the level of energy consumption since the quadrupling of petroleum prices, most projections of future consumption suggest that future growth in the demand for all energy and for electricity will be at lower than historical rates, even if sustained growth of economic activity is achieved in future.

The energy strategies adopted by the developed countries in recent years include attempts to achieve lower rates of growth of energy use by various means, including improvements in domestic, commercial and industrial design. In the long-run, these strategies, in conjunction with higher energy prices, may be expected to reduce the rate of increase in energy consumption. Although some studies have demonstrated the possibility that acceptable levels of economic activity may be achieved without substantial growth in electricity supplies, it appears that most countries expect their consumption of energy to increase in future, provided growth in economic activity also occurs.

6 ENERGY RESOURCES

In 1975, oil supplied 45 per cent of the world's primary energy (including that used to generate electricity), coal 30 per cent, and natural gas 18 per cent. Most of the rest was supplied by hydroelectric and geothermal sources, with nuclear energy contributing 1.3 per cent. These statistics exclude non-commercial sources of energy such as wood, charcoal, dung and animal power, which are important in many developing countries but of negligible importance in the developed countries.

Data on world consumption and world supplies of the principal forms of energy are shown in Table 3. In the table, the term 'reserves', applied to coal, oil, natural gas and uranium, refers to quantities, known to be present following geological exploration, which can be economically recovered with existing technology. For oil shale, bitumen rocks and thorium, which do not yet provide commercial energy, the 'reserves' figures are derived from assessments of the eventual economic capabilities of technologies now being developed.

Estimates of 'ultimately recoverable resources', which are many times higher than the 'reserves' estimates, include quantities of fuels not yet established, but expected to be present in existing proven areas, and speculations about other potential discoveries. Thus, not only is the existence of some of these resources uncertain, but considerable advances in technology may be required before they can be developed and used in environmentally acceptable ways. Nevertheless, it can be confidently expected that the quantity of recoverable reserves will increase, as it has done in the past, with further exploration and improvements in technology. Increases in fuel prices would contribute to this process.

Fossil fuels

Comparisons between recent levels of consumption of fossil fuels and the size of reserves are sometimes taken as indications of the urgency of reducing world consumption of these fuels. For example, if coal, oil and gas continue to be used at the same rates as in 1975, the reserves shown in Table 3 will be exhausted in approximately 206, 39 and 55 years respectively. But annual levels of consumption have risen considerably during the last 30 years, and are expected to continue to increase in the future. If this happens, those reserves will be exhausted more quickly.

However, while rates of consumption have increased, new exploration and technological change have continually added to both reserve and resource estimates in the past. Falls in the real prices of fossil fuels, such as petroleum and natural gas, up to the end of the 1960s tended to encourage increased consumption. ('Real price' means price from which the effects of inflation have been removed.) More recent events, particularly the increases in relative prices of these fuels, have served both to restrain the growth of consumption of fossil fuels and increase the rate of exploration for new supplies. The time period in which recoverable supplies of petroleum and natural gas will be exhausted remains very uncertain. On the basis of present knowledge, it seems possible that the contributions

Table 3

Estimated world consumption, recoverable reserves and resources of coal, oil and natural gas; 1975

	Consumption in 1975 (energy units: 10^{18} joules)	*Reserves (energy units: 10^{18} joules)*	*Reserves (physical units)*	*Ultimately recoverable resources (physical units)*
Coal (black and brown) .	73	15 080	665 x 10^9 tonnes	5 400–7 300 x 10^9 tonnes
Oil	112	4 360	110 x 10^9 m^3	210–300 x 10^9 m^3
Natural gas .	45	2 500	65 x 10^{12} m^3	80–170 x 10^{12} m^3
Oil shale . .	..	..	80 x 10^9 m^3	180–255 x 10^9 m^3
Bitumen rocks .	..	..	56 x 10^9 m^3	160–400 x 10^9 m^3
Uranium (thermal reactors).	3	(a)1 130	(b)2 700 x 10^3 tonnes	3 800–5 000 x 10^3 tonnes
Thorium . .	..	..	320 x 10^3 tonnes	2 000–2 800 x 10^3 tonnes

Sources: 1975 consumption: Information provided to the Commission by R. Krymm, Head of the Section for Economic Studies, Division of Nuclear Power and Reactors, IAEA.
Reserves and Resources: U.S. ERDA: *Creating energy choices for the future, 1976*
(a) Using conversion factor referred to in the text. Note that 1 tonne of uranium is equivalent to 1.30 short tons of U_3O_8.
(b) This figure is greater than that shown in Table 6, since it includes the estimated uranium reserves of the countries with centrally planned economies.

Table 4

Estimated Australian consumption, production and recoverable reserves of coal, oil and natural gas: 1975

	Consumption 1975	*Production 1975*	*Recoverable reserves (physical units)*	*Recoverable reserves (energy units: 10^{18} joules)*
Black coal . .	35 x 10^6 tonnes	67 x 10^6 tonnes	14 000 x 10^6 tonnes	360
Brown coal . .	27 x 10^6 tonnes	27 x 10^6 tonnes	12 000 x 10^6 tonnes	120
Oil	29 x 10^6 tonnes	20 x 10^6 tonnes	400 x 10^6 tonnes	15
Natural gas . .	5 x 10^9 m^3	5 x 10^9 m^3	850 x 10^9 m^3	33

Source: Information provided to the Commission by the Department of National Resources.
Recently revised estimates provided by the Joint Coal Board put recoverable reserves of black coal at 17 000 million tonnes and total resources, not all of which will be recoverable, at not less than 195 000 million tonnes.

of these two types of fuel to total energy requirements could be severely diminished by the end of this century.

Comparison of present consumption of coal with reserves shows its potential importance in meeting the world's energy requirements in the next century. Initially, its major impact will come in the replacement of oil and natural gas for the generation of electricity. Coal is already the most important fuel used in electricity production. The development of new ways of utilizing coal may reduce the relative consumption of oil and gas. A large research and development effort is under way in many parts of the world directed to achieving economical techniques for producing liquid and gaseous fuels from coal. New methods of producing heat from coal are also being investigated. One of these, fluidised bed combustion, is said to have advantages that include a great reduction in the amount of pollution produced. However, none of these newer technologies is expected to have a significant impact on the use of coal until after 1985.

There are some other fossil fuel resources which may help to relieve potential shortages. For example, oil shales and bitumen rocks (including tar sands) are potential sources of oil; estimated reserves and resources are shown in Table 3. Commercially viable technology for extracting the oil from these minerals has yet to be developed, so they can make no contribution to present energy production.

Uranium and thorium

Uranium and thorium are the other important non-renewable energy resources. Although uranium currently supplies only a small part of total energy consumption, it will become much more important if nuclear power programs proceed as currently projected. As the rate of consumption would grow rapidly under these circumstances, comparison of uranium consumption in 1975 with reserves would be very misleading and give no indication of the possible life of the reserves. The ability of uranium supplies to meet the anticipated demand is considered in some detail in Chapter 8.

It is also difficult to compare the energy content of uranium with that of fossil fuels. Such a comparison requires knowledge of the amount of energy that could be obtained from a tonne of uranium, which varies greatly according to the type of reactor in which it is used, the method of operation of the reactor and whether the fuel is recycled. One method of calculation, which may be appropriate for the present mix of thermal reactor types, without recycling of fuels, suggests that reserves of uranium would amount to about 5 per cent of presently estimated fossil fuel reserves. This method was used to derive the figures for energy value of uranium in Table 3.

If the fast breeder reactor eventually comes into commercial service, greater quantities of energy will be obtainable from uranium. But the factor by which these reactors will increase the effective energy content of uranium will not be known until they are established in commercial operation. The Commission was told that the theoretical maximum is about a 60 times increase, but that 30 times may be a more realistic estimate of what might be achieved in practice. On this basis, the effective energy content of known uranium reserves is potentially greater than that of known fossil fuel reserves.

The recent increases in the world prices for uranium have already resulted in increased exploration. Large increases in reserves can be expected, because many areas of the world with geological characteristics normally associated with the presence of uranium have not been explored.

Large amounts of energy could be obtained by breeding uranium-233 from thorium in reactors such as CANDUs, high temperature reactors and fast breeders, making thorium another potential nuclear energy source. As with the production of plutonium from uranium-238, each fuel charge in a reactor using this breeding process would have to include fissile material. In the early stages of such a program, this material would probably be uranium-235. It is not possible yet to give an assessment of the potential contribution of thorium to world energy supplies.

Australia's non-renewable energy resources

Table 4 shows Australia's current consumption, production and recoverable reserves of coal, oil and natural gas. It can be seen that Australia is particularly well placed, both absolutely and in relation to the rest of the world, for reserves of coal, but is less well endowed with oil and natural gas, particularly oil.

If Australia continued to consume the fuels at its 1975 rates, and did not export or import any, the black coal reserves shown in the table would last 400 years, the brown coal 444 years, the oil 14 years, and the natural gas 170 years. But some qualifications—in addition to those given earlier with the equivalent figures for world reserves—must be noted. For black coal, nearly half of Australia's present production is exported, and exports are likely to continue on a large scale. Including the present level of exports in the calculation shows the apparent life of the black coal reserves as 210 years. For oil, nearly one-third of Australia's consumption is at present imported. For natural gas, Australia may not be as well placed as the figure suggests, because consumption is growing rapidly as new pipelines connect more centres to supplies.

Australia's total reasonably assured uranium resources are put at about 350 000 tonnes, of which approximately 85 000 tonnes are estimated to be recoverable by the Ranger partners. The potential significance of these resources is considered in Chapters 8 and 9.

Alternative energy sources

Tables 3 and 4 make no reference to renewable sources of energy, those sources which are not based on finite stocks of mineral resources. Several witnesses put the proposition that urgent action should be taken to increase the world's use of renewable energy sources.

From the evidence presented, three main sources stand out as potential alternatives to fossil fuels and nuclear fission. They are direct and indirect solar energy, geothermal energy, and nuclear fusion. Fusion is not strictly a renewable energy source, but it is based on a resource so vast that for practical purposes it may be considered renewable.

Most renewable energy resources are derived directly or indirectly from solar energy. At present, the most important is hydro-power, which is an indirect form of solar energy. Together with geothermal and tidal energy, which made very small contributions, hydro-power supplied about 5.5 per cent of the world's energy in 1975, almost entirely as electricity. However, it is generally agreed that, even if all suitable sites for hydro-power production were developed, this source could supply only a small proportion of current and expected world energy consumption. Suitable sites for harnessing tidal energy are limited even more severely.

The potential contributions of solar energy, geothermal energy and fusion are now considered in turn.

Solar Energy

The Commission heard a great deal of evidence concerning solar energy. Since solar sources are renewable, there is no practical limit to the amount of energy that will be available from them in the future. However, there is a limit to the amount available to a country in any one year; this is determined by such factors as climate, the availability of suitable sites, competing uses for the land required and so on, as well as by the capabilities of the relevant technologies.

Many witnesses sought to show that there is no reason to believe the major technical problems associated with the further development of solar energy will not be solved. Its current commercial use for water heating to temperatures below 100° C was discussed, as were other direct applications including the heating and cooling of buildings, refrigeration and the production of heat at moderate temperatures (above 100° C) by means of heat-absorbing selective surfaces. Generation of electricity by solar furnaces and other steam-raising systems, by photo-voltaic devices, and by the utilisation of wind and differences in temperature between different layers of the ocean was also mentioned, as was the production of liquid fuels from photosynthetic materials including trees and field crops. Other witnesses took a less optimistic view of the technical prospects for solar energy. They described the large scale utilisation of solar energy for the generation of electricity as an enormous technological undertaking and stressed possible adverse environmental effects.

The economic feasibility of a greatly increased solar energy contribution is an even more contentious issue than its technical feasibility, and widely conflicting estimates of probable costs were given to the Commission. Unfortunately, most of the evidence on this topic, both from those taking an optimistic view of the potential of solar energy and from those taking a pessimistic view, was expressed in a form which makes it difficult to assess the likely contribution of the various solar energy technologies to future requirements.

The U.S. energy authority, ERDA, estimated in 1975 that an 'aggressive but potentially attainable' rate of introduction of solar energy could result in it providing some 7 per cent of that country's predicted energy requirements in the year 2000. A similar study by the U.K. Energy Technology Support Unit concluded that 5 to 7 per cent of British energy consumption in 2000 could be contributed by tidal and solar sources, including significant amounts of energy extracted from ocean waves. These estimates were based on economic growth continuing at rates slightly lower than those which have occurred since the Second World War. Lower rates of economic growth or reduced energy demand could increase the proportion of energy supplied by solar sources.

Many witnesses maintained that solar energy, in its various forms, could make a much larger contribution than these projections allow. They argued that the projections do not take into account possible changes in policies which, they claimed, currently discriminate against solar energy, for various historical and institutional reasons. If these policies were changed, it was suggested, a particular technology, once developed, could be implemented much more quickly. The advocates of solar energy welcomed increased funding for solar energy research in the last two years in many countries, notably the U.S.A. They called for further increases in financial support, which they believed would hasten the development of economically feasible solar energy technologies.

Geothermal energy

The reservoir of geothermal energy, the heat in the interior of the earth, is very large and, like solar energy, effectively inexhaustible. But there are very

24647/76—4

few places in the world where geothermal energy can be obtained naturally in its most useful form, steam. The World Energy Conference 1974 *Survey of Energy Resources* estimated that the upper limit of production of useful energy as electricity from this source would be 60 000 megawatts, which would represent only a very small part of world energy consumption.

The energy contained as heat in hot, dry rock inside the earth has much greater potential. So far, no way has been found to extract it, but consideration of possible extraction techniques has only recently begun. Japan expects geothermal sources to supply about 0.5 per cent of its energy requirements in 1985. An ERDA study suggests a possible geothermal contribution of 1.5 to 4 per cent of U.S. energy consumption by 2000. A British study projects a more modest 1 per cent contribution from geothermal sources to the U.K.'s energy supply at the turn of the century.

Fusion

Thermonuclear fusion is potentially a far more extensive source of energy than nuclear fission, even when possible output from breeder reactors is taken into account. A controlled fusion reaction has not yet been achieved. Evidence before the Commission indicates that there is optimism in some quarters that fusion power stations will become a reality. It is almost certain that this will not occur during this century.

It appears to be generally assumed that power generation from fusion will cause fewer environmental problems than is the case with nuclear fission, partly because the product of the fusion reaction is not radioactive. The Commission was told that induced radioactivity in a fusion reactor structure would be negligible within two years of a power station being shut down. Also, there would be no risk of the reactor getting out of control and no production of materials that could be used to build weapons. The main environmental problem would probably be leakage of tritium, a radioactive isotope of hydrogen. Present research is directed towards a fusion reaction requiring lithium, a mineral resource in finite supply. Eventually it is hoped to develop a system needing only deuterium, which is contained, in the form of heavy water, in all the earth's water; it is universally distributed and almost inexhaustible.

The recent OECD report *Energy Prospects to 1985* concluded that the scientific feasibility of controlled fusion might be demonstrated by the mid-1980s and that it is highly improbable that fusion will be applied to energy production before 2000. The development plan for fusion put forward by ERDA in the U.S. sets objectives of producing substantial quantities of electrical energy in two experimental reactors between 1985 and 1990, operating a commercial-scale demonstration reactor in 1997, and beginning to supply a fraction of the country's electrical energy demand after 2000.

Conclusions

While present indications are that world resources of oil and natural gas may be substantially depleted by the end of this century, this is not the case with coal. Relatively plentiful supplies of coal are available in many countries. Australia is one of the countries with large reserves of coal, capable of providing its own requirements and substantial exports for a very long time.

The potential contribution of uranium resources depends on the types of nuclear reactors used. It appears that if nuclear power programs proceed as currently predicted, uranium is capable of making a very significant contribution to total energy requirements, particularly if fast breeder reactors are eventually used as an important source of electricity.

Renewable energy sources, particularly solar energy in its various forms, seem likely to make important and growing contributions to total energy requirements, although the timing and extent of their contribution will depend heavily on the quantity of resources devoted to research and development. On present indications, their overall contribution may not be of great significance in the next decade. But they may make increasingly larger contributions thereafter, and become important by the end of the century. Demonstration of the feasibility of some of these technologies in the next decade may make a major commitment to nuclear fission energy unnecessary.

Figure 6

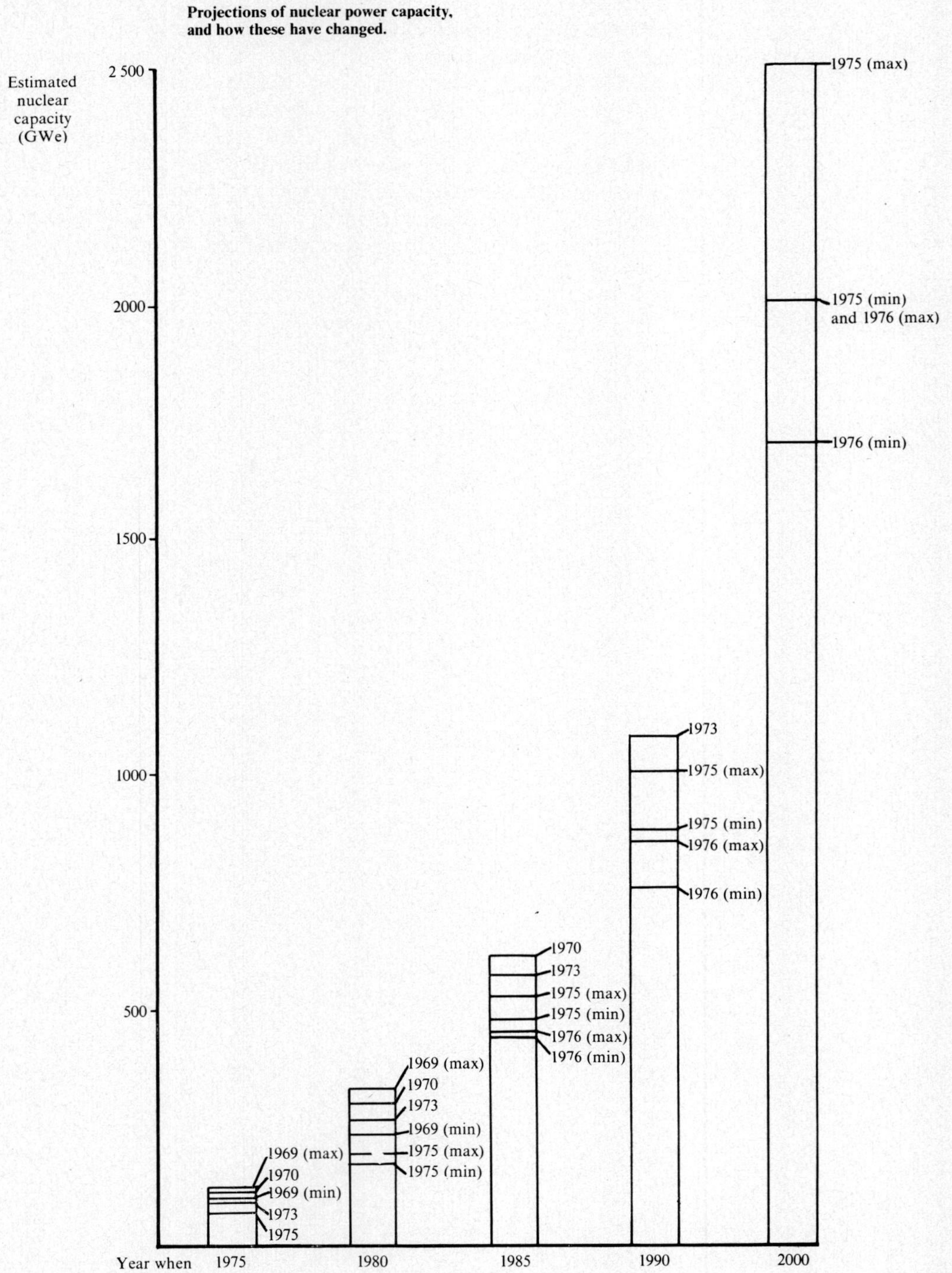

These projections, compiled by IAEA from information supplied by member governments, do not include estimates for countries with centrally planned economies. The chart shows that earlier projections tended to be considerably higher than more recent ones.

7 THE CONTRIBUTION OF NUCLEAR POWER TO WORLD ENERGY REQUIREMENTS

The total installed capacity of nuclear power stations has grown rapidly since 1970, from about 20 to more than 70 gigawatts electrical. It was anticipated earlier, particularly after the increases in oil prices in 1973, that a high rate of growth in commissioning of nuclear generating stations would continue. However, a marked reduction in the number of new orders occurred during 1974 and 1975, particularly in the United States. This reduction, and the deferment of other orders, can be expected to show up in a reduced rate of commissioning of new stations during the next decade.

Reductions in estimates of nuclear capacity likely to be in use in 1985 (see Figure 6) result mainly from downward revisions to national power supply plans. These revisions have been made necessary by the depressed demand for electricity since 1973, and by the excess generating capacity that has resulted from the commissioning of new plants at a time when demand has risen much more slowly than originally expected. The latest estimates from the International Atomic Energy Agency (IAEA), shown in Table 5, are broadly consistent with the most recent estimates made by the U.S. ERDA, also shown in Table 5.

Looking to the year 2000, estimates are inevitably much less certain. Table 5 gives a summary of the most recent estimates available to the Commission of possible nuclear generating capacity at the turn of the century, and Figure 6 shows how the estimates have been revised downwards. The estimates also demonstrate the wide range of possibilities which exist within the context of projections made by IAEA and ERDA. In fact, the range of possibilities is probably even wider, particularly after 1985, because of possible changes in rates of growth of economic activity and a variety of other considerations which may affect the rate of commissioning of nuclear power stations.

Opponents of uranium mining argued that projections of nuclear power capacity fail to take account of the demands on other energy sources resulting from a high rate of construction of nuclear plants. They suggested that energy analysis is the appropriate technique with which to consider these effects.

Energy analysis

Energy analysis is a recently developed approach to assessing investment policy decisions. It seeks to determine the costs and benefits of such decisions in energy terms rather than economic terms.

Energy analysis techniques have been used in a number of studies of the implications for total energy use of the growth of the nuclear power industry. These studies have generally begun by estimating the amount of energy needed to build a nuclear power station and operate it throughout its lifetime, together with the associated energy input at all other stages of the nuclear fuel cycle. Despite the difficulties involved in making these estimates, and the uncertainty of the data on which they are based, there is good agreement between a number of independent studies. The general conclusion is that a nuclear power station can be expected to provide ten to fifteen times as much energy as it consumes during a lifetime of about

Table 5

Estimates of Nuclear Generating Capacity (GWe), 1985 and 2000

	(1)	(2)	(3)	(4)	(5)	(6)	(7)
	OECD (1974)	*AUPF (1975)*	*Edison Electric Institute (1976)*	*AAEC (1976)*	*OECD-NEA/IAEA (1975)*	*ERDA (1976)*	*IAEA (1976)*
1985—							
United States of America . .	(b)275–358	216	185	189	205	160–185	(b)160
Western Europe .	175–227	197		182	193		185
Japan . . .	60–78	45	303	46	49	203–303	52
Other(a) . .	27–37	56		55	83		45
Total(a) . .	538–700	514	488	472	479–530	390–488	442
2000—							
United States of America . .	..	..	805	..	1 000	625–800	(b)700–800
Western Europe .	..	..		..	783		675–800
Japan . . .	..	..	1 451	..	157	1070–1450	(c)125–165
Other(a) . .	..	..		..	540		200–250
Total(a) . .	2 800–4 100	..	2 256	..	2 005–2 480	1 695–2 250	1 700–2 000

(a) excluding countries with centrally planned economies.
(b) including Canada.
(c) including Australia, New Zealand.

Column (1) OECD: *Energy prospects to 1985*, 1974.
(2) Evidence given to the Inquiry by the Australian Uranium Producers' Forum.
(3) Edison Electric Institute: *Nuclear fuels supply*, 1976.
(4) Evidence given to the Inquiry by the Australian Atomic Energy Commission.
(5) OECD-NEA/IAEA: *Uranium: Resources, production and demand*, 1975.
(6) Edward J. Hanrahan, U.S. ERDA: *Demand for uranium and separative work*. Paper given at Atomic Industrial Forum, Fuel Cycle Conference, Phoenix, Arizona, March 1976.
(7) Information provided to the Commission by R. Krymm, Head of the Section for Economic Studies, Division of Nuclear Power and Reactors, IAEA.

30 years. It pays back the energy used in its construction and initial fuelling either during its second year of operating at its maximum planned output or early in its third year.

Disagreements have arisen in what is termed dynamic energy analysis of nuclear power programs. This involves studying how energy is used and produced during the sequential construction of a number of nuclear power stations, and has usually been carried out by looking at the projected nuclear power programs of particular countries. It is generally assumed that their growth is exponential, which means that new nuclear capacity increases by a constant percentage each year. The total installed capacity increases at a steadily rising rate, precisely analogous to a sum of money earning compound interest, and this total doubles every so many years. In such a program, the energy used in building new stations may be an appreciable fraction of the energy being produced by those already completed. It is argued that, if the growth rate is very fast, energy used in construction activities may actually exceed the amount of energy produced. However, there is no doubt that, as soon as the rate of increase of a nuclear power construction program slows down, any deficit will be quickly repaid. In other words, there will be a transient energy deficit, analogous to a cash flow deficit incurred by a firm which has to borrow to finance capital expenditure.

Estimates have been made of the rate of exponential growth (or the time taken for nuclear capacity to double) at which such a deficit would start to be incurred. The estimates made in various studies differ, and disagreements have arisen as to whether national nuclear power programs have grown or will grow quickly enough to incur a deficit.

The disagreement among energy analysts arises mainly from differences in conventions adopted for conversions between electrical and thermal energy. Depending on the conventions used, the studies indicate that a net energy deficit will be incurred by a nuclear power program with a doubling time varying from one to three years.

A doubling time of three years implies an annual growth rate of 26 per cent. After an initial rapid build-up, the U.S.A. certainly appears to have entered a period of slower nuclear growth with a projected doubling time of more than 4 years up to 1985. But some other countries, including the Federal Republic of Germany and Japan, have projected doubling times up to 1985 of about three years, so there seems to be some possibility that these programs will incur an energy deficit during the next few years.

The application of energy analysis shows that a rapid increase in nuclear capacity may result in a temporary increase in the use of other energy sources. However, this deficit will be made up in a relatively short space of time if the rate of installation of nuclear capacity ceases to increase exponentially.

It seems that the application of energy analysis could show up constraints on a country's program which might not be apparent from traditional economic analysis. This would occur if the economic analysis omitted the effects on costs and overall economic activity of a reduction, during a rapidly growing nuclear program, in the electrical energy available to the economy outside the energy sector. Even in periods when the nuclear program did not incur an energy deficit, there could be economic effects which might not be predicted by traditional economic analysis.

Economics of nuclear and coal-fired electricity generation in economically advanced countries

The growing nuclear contribution in national electricity supply projections is generally the response to a belief that nuclear power has an economic advantage over alternative methods of electricity generation, at least for meeting the continuous (base) load placed on large electrical grid systems. Opponents of nuclear power challenged the contention that nuclear facilities hold a substantial economic advantage over coal-fired plants for these base-load operations, and argued that coal remains a cheaper source of electricity than nuclear power in many locations. They also argued that the costs of nuclear generation have increased more rapidly in recent years than the costs of coal-fired generation.

The average cost per unit of electricity depends principally on the type and size of generating equipment used, the type and cost of fuel, the efficiency of the generating plant, and the extent to which demand fluctuates during a day and over a year. These factors can vary significantly between different parts of a country, so little reliance can be placed on unit cost data based on national averages as a guide to decisions on the most economic type of plant in any area. In economically advanced countries, nuclear stations generally have higher capital costs and lower running costs, per unit of electricity generated, than stations burning fossil fuels. Any economic advantage held by nuclear power stations therefore must rest on the operating costs being low enough to offset higher capital charges.

Capital costs Large increases in the capital costs of building and commissioning nuclear stations in recent years have been documented in various studies. These studies show that rising construction costs and lengthening of the lead time from the planning to commissioning stages have been the principal reasons for these increases. Some witnesses claimed the increases were likely to reduce any economic advantage held by nuclear stations over coal-fired alternatives, and thus further adversely affect nuclear power programs. Others rejected this argument, pointing to evidence that similar increases had occurred in the capital costs of coal-fired stations, particularly in relation to requirements for scrubbers to reduce sulphur emissions.

Data compiled by the Head of the Section for Economic Studies in the IAEA's Division of Nuclear Power and Reactors confirm the weight of evidence that the relative advantage in capital costs held by coal-fired stations has not diminished, despite the requirement for scrubbers. Any overall advantage held by nuclear stations must continue to rely on lower operating costs.

Some witnesses suggested that the heavy capital costs involved in proposed nuclear power programs might place such a burden on available resources as to stifle investment in other desirable areas. The OECD's study *Energy Prospects to 1985* shows that nuclear power programs are expected to absorb increasing proportions of available capital funds in each OECD country up to 1990. The estimates vary considerably between countries, and the proportions will, of course, depend on the nuclear capacity actually installed. The study notes that the total capital investment predicted for all energy programs in the next decade would represent a very large shift of resources into the energy sector, and that these programs may conflict with other economic objectives.

The U.S. Federal Energy Administration's (FEA's) *1976 National Energy Outlook* says the cost increases in recent years, and the failure to adjust electricity prices quickly to the new cost situation, were responsible for much of the difficulty experienced by U.S. electricity authorities in financing their nuclear power programs. Internal funds available for financing new investment were severely restricted, making it necessary to raise large amounts of capital from market sources. The report shows that borrowings by U.S. electricity utilities absorbed a substantially increased proportion of savings in the United States in recent years. Because of increases in interest rates and the need to offer higher returns on equity issues, obtaining the necessary finance has been difficult and the cost of electricity has risen further. The increases in capital costs and lead times for new stations coincided with the economic recession and a relative shortage of available capital funds. The report says the electric utilities may find it difficult to achieve the level of financing required for their nuclear programs, particularly because it may prove difficult to make new equity issues that will prove attractive in the capital market.

The evidence suggests that similar problems have been experienced elsewhere. It appears that electricity authorities generally may experience difficulties in raising finance for heavy capital expenditures in the decade ahead. If this proves to be the case, financing costs are likely to rise and, consequently, so are the prices charged to electricity consumers. Since nuclear plants are more capital-intensive than coal-fired plants, it seems probable that the relative cost position of nuclear plants may be adversely affected by these financing difficulties.

Capacity factors

A power station's capacity factor is the amount of energy it actually produces in a given period divided by the amount it would have produced if it had operated at full capacity throughout the period. Ultimate unit costs of electricity generation depend to a considerable degree on the capacity factors achieved; lower capacity factors generally result in higher unit costs. Whenever a power station reduces its output or closes down, whether in response to varying demand for electricity or to allow for maintenance which becomes necessary because of plant failures or for any other reason, its capacity factor falls.

Because there is no economical way of storing great quantities of electricity, it is generally necessary to have sufficient generating capacity to match the maximum demands made on supply grids. Since large nuclear power stations generally have lower operating costs than large fossil-fuelled stations, the usual role of the nuclear stations is to supply base-load electricity rather than the additional power needed at periods of intermediate and peak demand.

Comparisons of capacity factor are meaningful only if they are between power stations with similar roles in their respective grids. Evidence before the Commission shows that many existing base-load nuclear plants have not achieved the capacity factors originally anticipated, and that their average capacity factors have frequently been below those achieved by coal-fired plants used to supply base-load electricity. Other evidence shows substantial declines in capacity factors with increases in the size of nuclear plants. Large-scale fossil-fuelled plants also have generally achieved significantly lower capacity factors than their smaller-scale counterparts.

In summary, both large nuclear and large coal-fired stations used for base-load electricity generation have achieved lower capacity factors than smaller-scale plants fuelled by coal, oil or natural gas. This situation calls into question the belief that very large generating units will produce electricity at lower cost than smaller units, which in turn suggests that smaller-scale coal-fired generating plants may prove to be more economic than large coal-fired or nuclear generating units. The evidence also indicates that small-scale nuclear plants are unlikely to be economic when compared with coal-fired units of the same size.

Fuel costs

In computing the overall costs of electricity generation, capital costs, including interest and depreciation charges, have to be allocated to the units of electricity generated, and then added to the running costs. The main item among the running costs is the cost of fuel, and the rises in fuel prices in recent years have increased the unit costs of producing electricity. Oil prices now make construction of new oil-fired generating plants uneconomic, except in a few places where alternative fuels would continue to be more expensive. Since some existing oil-fired plants can be converted to coal-burning units, a number of generating authorities have reduced their costs, and demand for petroleum, by making the conversion.

Relative costs of nuclear fuel and coal will be an important consideration in determining the comparative economic merit of nuclear and coal-fired electricity generation in future. Earlier expectations that the real cost of nuclear fuel would decline through time have not been realised. The costs associated with all parts of the nuclear fuel cycle have risen significantly in recent years, and an expectation exists that further increases will occur in most parts of the cycle in the future.

Up to the present, the nuclear fuel enrichment costs paid by most electricity generating authorities have been based on prices charged by U.S. ERDA, but these prices cannot be taken as indicative of the real cost of providing enrichment services for future nuclear fuel requirements. Unless governments continue to subsidise enrichment facilities, prices will be set to reflect all capital and operating costs. There is evidence that the real costs of enrichment, per unit of electricity generated, may be at least double recent ERDA prices.

As noted in Chapter 4, no commercial plants for recovering uranium and plutonium from spent oxide fuel, with the object of recycling it for further use, are in operation yet. The profitable operation of such facilities will depend on their ability to cover capital and operating costs from the sale of recycled fuel to electricity authorities. The evidence suggests that reprocessing may not be a profitable operation at the average price of uranium which seems likely to apply in future, and there is therefore considerable doubt as to whether it will prove to be commercially attractive without government support. If reprocessing is carried out even though it is not a profitable operation, there will be a further increase in the real cost of electricity generated by nuclear power stations, whether or not the prices actually charged for electricity include such costs.

Although available estimates of the costs of waste disposal represent only a small part of the anticipated overall costs of nuclear electricity generation, it is evident that these costs are subject to a considerable degree of uncertainty. This uncertainty will not be resolved until final solutions to waste disposal problems are found and the costs accurately assessed. The cost of disposing of reactors at the end of their life must also be considered. Some evidence suggests that this may be about five per cent of the initial capital cost of the reactors.

It is conceivable that the real costs of reprocessing and waste disposal or storage will not be met by electricity charges, but will be subsidised by governments. In this case, the 'polluter pays principle' will not be followed, and the competitive position of nuclear power compared with alternative energy sources will be artificially improved.

On balance, the evidence suggests that recent and expected future changes in the real cost of nuclear fuels will add to the real cost of electricity generated by nuclear stations. In considering the likely effects of these increases on the relative economic merits of nuclear power generation, account must also be taken of expected changes in the real cost of coal suitable for electricity generation. Evidence given to the Commission stressed the importance of the availability and cost of coal, including the cost of transport, as a major determinant of the cost of coal-fired electricity. Since the Second World War, coal costs have varied according to the location of the coal and whether it could be mined by open-cut methods. The average costs of coal obtained from open-cut operations are generally much lower than those from underground mining. Following the increases in oil prices in 1973, coal prices increased substantially as greater demands were made on existing supply sources.

Technical change has been an important source of reductions in real costs of coal in many areas where coal production has increased since the Second World War, but the incentive to carry out research and development into mining and use of coal has been restricted because of the competition provided by cheap supplies of petroleum. Now that the increase in petroleum prices has made coal much more economically attractive, research and development expenditures are being substantially increased and there seems little doubt that technical advances will be made so that greater advantage can be taken of the world's coal resources in the future. Similar technical advances may help to reduce the real costs of nuclear fuels.

Potential for nuclear power in principal developed countries

Substantial increases in the relative price of petroleum in recent years have had a major influence on energy policies in the developed countries. Much more importance has been attached to the objective of reducing dependence on imported supplies of energy, and to achieve this aim a good deal of attention has been devoted to energy conservation, the development of indigenous energy sources, and the diversification of imports, by type and source, to provide greater security in the event that supplies are curtailed. The following review of evidence relating to the major users of nuclear power indicates the extent to which these considerations appear to have influenced recent energy developments in these countries.

U.S.A.

The United States Federal Government is at present committed to further development of nuclear power, principally as a means of reducing dependence on petroleum imports. However, the U.S.A. has at present substantial excess of generating capacity over requirements, and it is not anticipated that this excess will be eliminated for some time.

Issuing of new operating licences by the U.S. Nuclear Regulatory Commission for all parts of the nuclear fuel cycle is at present interrupted due to a court determination in the District of Columbia regarding the requirement to demonstrate the solution of certain environmental problems. The N.R.C. believes this delay will not persist more than a few months.

Based on average costs for the whole country, the U.S. FEA's *1976 National Energy Outlook,* published recently, concludes that: 'Nuclear energy is the cheapest source of base-load electric power, although not much cheaper than coal'. But it also indicates that recent rises in capital costs have increased nuclear generation costs more than those of coal-fired stations. And it observes at one point that:

> Nuclear and coal generation costs are close. The delivered price of coal varies over a wide enough range that in some regions coal plants may generate electricity for less cost than do nuclear plants. Indeed, coal and nuclear plants are close enough that they might be considered the same, given the uncertainty associated with the estimates.

The report adds that there appears to be a nuclear/coal trade-off 'where the economic criteria may make little difference and where the decision between the two or the proper mix of the two may depend, therefore, on an assessment of the environmental and social costs and risks associated with them '. Elsewhere the report indicates that, although there are short-run limitations on the supply of coal, these may not apply in the long-run.

Japan Japan, possessed of relatively few of the natural resources required by a modern industrial society, is highly dependent for its economic well-being on the efficient production of manufactured goods. Its economy depends heavily on the availability of energy on a reliable, economic basis. Imports from other countries meet most of its energy requirements.

The evidence indicates that Japan is at present committed to an energy strategy involving the establishment of long-term arrangements for the supply of energy resources, which will provide the basis for a rate of increase in economic activity and energy use somewhat lower than post-war trends up to 1973.

Although nuclear energy now supplies only a small proportion of its total energy requirements, Japan expects that continued growth in the level of economic activity will require an expansion of its nuclear power industry. However, difficulties are already being met in finding publicly acceptable sites for nuclear power plants, and this could be a limiting factor in its nuclear development. A substantial increase in imports of coal and liquid natural gas for electricity generation is also expected in the next decade.

Replacement of part of Japan's proposed nuclear expansion program with an amplified coal-fired power station program, combined with greater emphasis on energy conservation, appears feasible. But the extent to which this could be achieved, or would be acceptable to Japan itself, cannot be assessed from the evidence. Base-load electricity generated by stations fuelled with imported coal would probably be more expensive than that provided by nuclear stations in many cases. However, the evidence indicates that the average capacity factor of some existing nuclear stations in Japan has been low, so the cost of electricity produced by them may be relatively high.

Western Europe Most of the nuclear power capacity in Western Europe is likely to be installed in member countries of the European Economic Community. The Community's full members are the U.K., France, the Netherlands, the Federal Republic of Germany, Belgium, Ireland, Denmark, Luxembourg and Italy. The size of the existing nuclear industry varies from nation to nation, as does the expected future degree of reliance on nuclear power.

The U.K.'s operating nuclear power stations account for 12 per cent of the country's electrical power capacity, and the U.K. has fully developed uranium enrichment and fuel reprocessing technologies. Due to recent levels of demand falling considerably below earlier expectations, it at present has a surplus of generating capacity, suggesting that its needs for additional power stations—nuclear or otherwise—will be modest in the next decade. The U.K. has very substantial coal reserves relative to its needs. It also has the North Sea oil and gas reserves, which are anticipated to meet a large proportion of its energy requirements for some time. Consequently, the U.K. is not expected to require additional nuclear capacity for perhaps a decade.

The Federal Republic of Germany has seven operating commercial nuclear power stations. It has developed an enrichment technology to the pilot plant stage, and prototype reprocessing plants are operating. Nuclear installations now provide about 5 per cent of West Germany's electricity, and the evidence indicates that a maximum of 45 per cent of the total may be provided by nuclear stations in 1985 and possibly 75 per cent by the end of the century. If these projections are realised, there will be continuing increases in uranium requirements.

The Federal Republic of Germany has significant coal resources, which are currently being exploited. However, like other countries, it is heavily dependent on imported oil and is anxious to remove a large part of this dependence by increasing its production of energy from other sources. Known indigenous supplies of uranium are relatively small. The F.R.G. is associated with uranium exploration and development ventures in other countries, and hopes to meet 60 per cent of its requirements from its own deposits or those in which it has an interest.

In summary, the evidence indicates that the Federal Republic of Germany believes nuclear power will provide the most economic means of meeting most of its additional base-load electricity requirements in the next decade.

All other countries of the EEC currently expect increases in energy demand to continue through the next decade provided levels of economic activity continue to increase. An increasing proportion of electricity requirements is expected to be met by nuclear stations in most of them.

Potential for nuclear power in developing countries

Although total use of energy and electricity in developing countries has risen at average rates greater than those in developed countries, the developing countries together account for only a small proportion of the energy and electricity consumed in the world. Moreover, consumption of energy is heavily concentrated in the more advanced of the developing countries. In 1974, the last year for which complete statistics are available, India, Brazil, Mexico and Argentina accounted for about 43 per cent of the total consumption of electricity in the developing countries.

The developing countries account for only a small proportion of nuclear generating capacity currently in use, installed capacity amounting to only about 1 gigawatt electrical at 31 December 1975. Projections submitted to the Commission suggest that nuclear energy may make a more important contribution to the needs of some of these countries in the decades ahead. However, projections by IAEA of total nuclear capacity in developing countries in the year 2000 have been revised downwards recently from 400 to 200-250 gigawatts electrical. If the most recent predictions are fulfilled, these countries could have about 10 per cent of world nuclear capacity (excluding capacity in the centrally planned economies) in 1985 and 2000.

However, a number of witnesses argued that nuclear power has a limited contribution to make in most developing countries because it is not suited to their needs. Even if the demand for electricity does grow comparatively rapidly in these countries in future, it does not necessarily follow that there will be a rapid development of extensive grid systems capable of supporting large power generating units. It was suggested that these countries frequently need generating stations with a capacity of about 100 megawatts electrical, for which nuclear generation is generally uneconomic. Another point made was that, in many cases where distribution grids exist, they supply electricity for a small, affluent group of people living in cities rather than the rural masses. Consequently, the people most in need of economic assistance would be unlikely to receive any direct benefit from the provision of additional electricity. It was also suggested that earlier forecasts of nuclear requirements ignored the substantial coal reserves available in many developing countries.

Some witnesses who supported these views went on to argue that, if developing countries became more dependent on nuclear power, they would need to rely more heavily on imported technology, materials and equipment than at present. They suggested that technically simple, small scale, cheap and relatively labour-intensive energy technologies, based largely on equipment which the developing countries can manufacture themselves, would be more suitable for the developing countries. It was submitted that if oil is unavailable, 'soft' technologies, such as solar and wind energy, may come nearest to satisfying these criteria. It was also pointed out that most of the poorer countries are located in the tropics, where abundant solar energy falls.

A number of witnesses, while agreeing that nuclear power would be of little direct benefit to the developing countries, said that the use of nuclear energy in the developed countries could indirectly help the developing countries (especially those which do not produce oil themselves), principally by increasing the amount of oil available for their use. While a more rapid depletion of the world's oil resources will in time affect all nations, it is evident that it would have the greatest effect on the poorer developing countries, which would be least able to afford the higher prices that are likely to be charged for oil as supplies become progressively more difficult to obtain.

It was also submitted that nuclear power would indirectly assist developing countries by contributing to economic growth in the developed countries, thus stimulating the volume of potential exports from the developing countries and making additional economic assistance from the developed countries more readily available. However, the latter view was strongly contested. It was submitted that development strategies relying on private investment and government aid, while they may have prevented the differences between poor and rich from becoming even wider, have generally failed to reduce the disparity in incomes between rich and poor countries. It was said that new policies are necessary; that these should place particular emphasis on deliberate measures to expand trade and increase the prices developing countries receive for their exports and that the implementation of such measures should not depend on the level of prosperity in developed countries. Some witnesses also argued that the very large demands for capital in the developed countries stemming from capital-intensive energy programs including nuclear power may serve to reduce the flow of investment capital to the developing countries.

These witnesses also pointed out that plans for the export of uranium from Australia envisage that most of it would be sold to the developed countries of Western Europe, North America and Japan. They argued that Australia would thereby be directly helping the countries which already consume a disproportionate share of the world's resources. Such a policy would not reduce the unequal access to, and consumption of, resources between developed and developing countries, which is one of the most important causes of international tension today. Hence these witnesses concluded that, by exporting uranium, Australia would not be reducing international tensions arising from the uneven distribution of resources, as some of the witnesses suggested, but would, if anything, help to increase such tensions.

Conclusions

There is an expectation of an increasing, but not total, reliance on nuclear energy to cope with growth in base-load electricity demand in the countries likely to be the principal potential purchasers of Australian uranium. However, there are also indications that existing plans for extensions of nuclear power in economically advanced countries may be based on over-optimistic predictions of the cost of electricity from nuclear power stations. Rising capital costs and difficulties in financing the heavier investment required are affecting nuclear plants more severely than coal-fired stations. Expected increases in nuclear fuel cycle costs will also adversely affect the competitive position of nuclear stations. The Executive Director of the OECD's International Energy Agency recently stated that, 'because of the steep increases in both capital costs and fuel costs, it is not certain that nuclear power will continue to be cheaper than electricity generated by fossil fuels'.

A further decline in the competitive position of nuclear power stations may cause further reductions in their rate of construction. On the other hand, the expressed desire of the developed countries to become more independent of imported supplies of energy may encourage greater reliance on nuclear power. It appears that most of the economically advanced countries will proceed with the building of further stations, the rate of construction depending heavily on the rate of growth in demand for electricity, but also on the relative cost of nuclear power and on the desire for limiting dependence on imported fuels or on one energy source. Electricity demand will be closely related to the future rate of growth of economic activity, predictions of which involve great uncertainty.

Where adequate and relatively cheap supplies of suitable coal are available, as in much of Australia, nuclear power remains at an economic disadvantage, even for base-load requirements. Large quantities of coal are available in many parts of the world, including the United States and some Western European countries. New coal-fired stations, rather than increased nuclear capacity, could be built to cope with increased demand for electricity, at least until other sources of energy are more fully developed. In this case the cost of generating and supplying electricity would probably rise in many areas, the extent of any increase depending on the relative costs of nuclear and coal-fired generation, which would vary between countries and regions. However, the evidence indicates that use of coal rather than further extensions of nuclear power would not, in general, cause very substantial overall increases in the cost of electricity or in the general level of prices.

Supplying the necessary coal would, of course, require provision of adequate mining and transport facilities. Also, measures would have to be taken to protect the environment, as they would have to be if nuclear capacity was increased instead. Some countries, notably Japan, would have to rely on greater imports than currently anticipated for an increased coal-fired electricity generating program, and this may be considered undesirable for strategic reasons. In any case, substantial government support would be required in order to achieve much greater reliance on coal.

It is evident, from the information available, that existing estimates of the cost of electricity generation include costs associated with the prevention of undesirable environmental effects, at least to the extent that such costs have to be met by electricity generating authorities. It is also apparent, however, that some costs associated with the nuclear fuel cycle, such as the total costs of uranium enrichment, are not being met by electricity consumers. There is a strong prospect that a similar outcome will occur in relation to the costs of reprocessing spent fuel. On a wider scale, it is also evident that possible 'external' costs arising from sources such as environmental effects of the operation of nuclear reactors, the reprocessing of fuel, the disposal of wastes and the possible diversion of plutonium to production of weapons, as discussed in later chapters, make it extremely unlikely that 'the polluter pays' principle will be fully applied and that all such costs will be internalised. As will be evident from the discussion in Chapter 10, some of the most important 'external' effects associated with the use of nuclear fuels continue for extremely long periods of time. As a consequence, it appears likely that governments will have to take action to ensure that these costs are taken into account in reaching final decisions about the overall desirability of nuclear power as a source of energy.

As discussed in Chapter 11, there are also significant 'external' effects stemming from the use of fossil fuels for electricity generation. Those associated with the mining and use of coal are particularly important. Such factors should also be taken into account in decisions relating to energy use. It was also suggested that a high rate of use of fossil fuels, such as coal, for electricity generation would deprive future generations of the use of those fuels for chemical and metallurgical purposes. However, in view of the great quantities of coal available to the world, any world-wide shortage of this resource seems unlikely for many decades.

The evidence also points to the conclusion that, while some of the more advanced developing countries may proceed with plans to install nuclear capacity during the remainder of the century, nuclear power is unlikely to contribute on a large scale to the energy needs of the less affluent countries. Nor does it appear that the further development of nuclear power in economically advanced countries will make any significant difference to the ability or the willingness of those countries to assist less affluent countries. The possibility that Australia could be of greater long-term assistance to the rest of the world, and in particular to less affluent countries, by participating in international efforts to develop those forms of solar energy technology most suited to the needs of developing countries appears worthy of serious consideration. This matter is discussed further in Chapter 15.

8 URANIUM: SUPPLY AND DEMAND

One of the main tasks of the Commission is to examine the likely effects of a decision to export Australian uranium and of alternative decisions to withhold supplies temporarily or permanently. A key question is the extent to which world uranium reserves are available in the quantities estimated to be required for electricity generation. The availability of uranium from other countries will influence the extent to which Australian uranium is likely to find markets in the rest of the world if exports are permitted. It will also be important in determining the probable effects of alternative decisions upon countries with nuclear power programs.

In this chapter, we relate information concerning the extent of uranium reserves to the anticipated uranium requirements for nuclear power generation over the rest of this century. The extent to which buyers of uranium will wish to make purchase arrangements with potential suppliers in the immediate future is also considered.

Uranium prices

Large variations have occurred in uranium prices in the last two decades. During the 1950s, when many uranium mines were being developed in the U.S.A. and other countries, the average price was approximately US$11 per pound of U_3O_8 in yellowcake. With the slow increases in demand recorded in the 1960s, prices fell to much lower levels. In this period, the principal buyer was the U.S. Atomic Energy Commission, which purchased uranium for its stockpile. Production of uranium generally exceeded demand, putting downward pressure on prices and resulting in large build-ups in stocks. The price fell to below US$5 per pound of U_3O_8 in 1972.

Demand for uranium did not increase significantly until the advent of substantially higher oil prices late in 1973. As surplus inventories were disposed of and existing production capacities more fully utilised, prices began to increase sharply. The average price of U_3O_8 in yellowcake for early delivery increased to about US$15 per pound in new contracts made at the end of 1974, to about US$35 per pound in new contracts made at the end of 1975, and to US$40 per pound in new contracts made in the first half of 1976. Prices paid under previously existing contracts for deliveries in these periods were, however, much lower, averaging less than US$11 per pound of U_3O_8 in yellowcake. Recent price increases partly reflected general inflationary conditions, but stemmed mainly from the desire of utilities to cover their more immediate requirements at a time when the rate of uranium production could not be quickly increased.

Substantial numbers of contracts for future deliveries of uranium have been negotiated since 1974. The Commission was told that some of these contracts contain provisions for prices to be based on the market price applying at the time of delivery with a formula which provides for cost increases due to inflation to be recovered. Other contracts do not refer to market prices, but include a contract price which is to be adjusted by formula to cover cost increases. Actual prices paid in future years may be substantially above average levels of recent prices, but 'real' prices, adjusted to remove the effects of inflation, will not necessarily rise.

There are indications that the effects on the market of attempts to secure supplies of uranium up to 1979 are subsiding, and that more attention

24647/76—5

is being devoted to requirements into the 1980s. However, uncertainties relating to the rate of growth of nuclear power generating capacity may, on future occasions, cause large fluctuations in the demand for uranium. As demand is unlikely to be responsive to short-run price changes, stocks of yellowcake are likely to fluctuate and uranium prices to react violently, as they have done in the past. Provisions for deliveries under contracts may have to be renegotiated, and production adjusted accordingly. It is thus possible that there will be periods of time during which mining capacity will be idle and uranium prices depressed. The discussion which follows concentrates on the likely long-run trends in supply of and demand for uranium.

Uranium resources and requirements

A summary of the most recent data relating to the availability of uranium is set out in Table 6. The data are arranged according to the classifications used in the most recent OECD-NEA/IAEA report on the subject. The category 'reasonably assured resources' refers to uranium, occurring in known ore deposits, recoverable with currently proven mining and processing technology. The category 'estimated additional resources' refers to uranium which is expected to be discoverable in unexplored extensions of known deposits or in undiscovered deposits in known uranium districts.

The term 'reserves' is applied only to reasonably assured resources currently included in the cost category below US$15 per pound of U_3O_8 in yellowcake. Cost categories are based on estimates which include the direct cost of mining, milling and extraction, and the writing-off of capital used in providing and maintaining production units. All profits and exploration costs are excluded.

Future uranium requirements for nuclear programs will depend mainly on the rate of growth of nuclear power generating capacity, and the fuel requirements and operating characteristics of the reactors used. They will also be influenced by the amount of uranium-235 remaining in depleted uranium from enrichment plants (the tails assay), and the quantities of uranium and plutonium recycled. A summary of the various estimates of uranium requirements presented to the Commission is contained in Table 7. The table also sets out the assumptions made about tails assays and recycling in deriving each set of estimates.

The downward revisions over the past two years in anticipated additions to nuclear capacity have led to reductions in estimated uranium requirements. As shown in the last column of Table 7, cumulative uranium requirements, as estimated by ERDA, to provide the fuel needed in the United States, would total 145 000 to 198 000 tonnes of uranium up to 1985 and approximately 1 million tonnes up to the end of the century. Total requirements for all countries (excluding the centrally planned economies) up to 1985 would be between 401 000 and 476 000 tonnes, and about five times that amount would be required up to the end of the century.

The differences between these estimates and others provided to the Commission are due mainly to the downward revisions in orders for nuclear capacity reflected in the latest ERDA projections. There are also some differences relating to tails assays and the timing of recycling and reprocessing. Because of the long lead times involved in planning and commissioning new nuclear stations, the maximum amount of installed capacity in 1985 is unlikely to exceed ERDA's lower figure in Table 5.

Slower growth in economic activity and a worsening of the cost position of nuclear power could cause further deferments in the commissioning of nuclear capacity, with consequent reductions in the use of uranium. However, it is unlikely that higher rates of economic growth, reduced costs or the solution of environmental difficulties could cause any significant acceleration in nuclear power programs up to 1985. In the long term, the availability of petroleum supplies will be a major factor in determining the amount of nuclear capacity installed. Recent and projected increases in the real costs of electricity generated by nuclear power stations, or lower-than-predicted growth in economic activity, may lead to substantial further reductions in nuclear programs. Clearly the estimates of nuclear capacity for the remainder of the century are subject to great uncertainty, and so are the estimates of uranium requirements. Uncertainty about uranium requirements also arises from the fact that future decisions about tails assays, recycling and the introduction of fast breeder reactors cannot be accurately predicted.

Uranium reserves in the category below US$15 per pound of U_3O_8, available in countries other than the centrally planned economies, clearly exceed estimated uranium requirements up to 1985 by a considerable margin. The evidence suggests that more than half of the currently defined reserves available outside Australia may be needed to provide fuel for electricity generation in this period. Looking further ahead, the existing estimates for requirements up to the year 2000 could not be met from reasonably assured resources in the cost categories up to US$30 per pound, but would require exploration, development and use of a large amount of uranium currently in the 'estimated additional resources' category. The inclusion or exclusion of Australian supplies, as shown in Table 6, makes no substantial difference to this conclusion, since Australian deposits account for only about 9 per cent of the total of all reasonably assured and estimated additional resources of uranium up to US$30 per pound.

This conclusion is consistent with the view expressed by many witnesses that the non-availability of Australian supplies would not prevent other countries from proceeding with their nuclear power programs. Many countries have taken only preliminary steps to explore and prove their uranium resources, mainly because of the low price of uranium before 1974. In view of price increases since then and the interest shown more recently by Western European, Japanese, and United States electricity authorities in obtaining uranium supplies, it seems certain that there will in the future be further substantial increases in proven reserves and other resources. Similar increases may be expected in Australia if exploration for uranium continues.

A substantial increase in international exchanges of uranium is likely in the coming decades if nuclear power programs go ahead on the scale currently projected. Even if, as seems probable, the United States relies principally on its own supplies, European and Japanese requirements can only be met by substantial imports. These countries already have arrangements to import from other areas, but are interested in obtaining further supplies from Australia and elsewhere.

Uranium production capacity

Estimates of the amounts of uranium likely to be required by U.S. electricity generating utilities in the next decade, which have not already been contracted for, depend to some extent on assumptions about U.S. electricity demand. But they also depend on assumptions made about additional uranium supplies likely to be available from U.S. sources and the stocks of uranium which utilities may wish to hold in future. There is good agreement among the various recent estimates of total cumulative U.S. uranium

Table 6

Uranium Resources (thousand tonnes uranium)

	Reasonably assured resources						Estimated additional resources		Total reasonably assured and estimated additional resources	
	Up to US$15 per pound U_3O_8		$15–$30 per pound U_3O_8		Total up to $30 per pound U_3O_8		Up to US$30 per pound U_3O_8			
		Percentage of total		Percentage of total		Percentage of total		Percentage of total		Percentage of total
North America—										
Canada	145	12.6	28	3.2	173	8.6	605	33.1	778	20.2
United States of America	331	28.8	(g)269	30.9	(g)600	29.7	815	44.6	(g)1 415	36.7
Mexico	5	0.4	1	0.1	6	0.3	..	..	6	0.2
Total	481	41.8	298	34.2	779	38.6	1 420	77.7	2 199	57.1
Western Europe(a)	61	5.3	(h)426	48.9	(h)487	24.1	181	9.9	(h)668	17.3
Africa—										
Algeria	28	2.4	..	..	28	1.4	..	..	28	0.7
Gabon	20	1.7	..	..	20	1.0	10	0.5	30	0.8
Niger	40	3.5	10	1.1	50	2.5	30	1.6	80	2.1
South Africa(b)	186	16.2	90	10.3	276	13.6	74	4.1	350	9.1
Total(c)	274	23.8	100	11.5	374	18.5	114	6.2	488	12.7
South America(d)	19	1.7	12	1.4	31	1.5	48	2.6	79	2.1
Asia(e)	4	0.3	35	4.0	39	1.9	23	1.3	62	1.6
Total (excluding Australia)(f)	839	72.9	871	100.0	1 710	84.6	1 786	97.7	3 496	90.8

Australia—										
Ranger	85	..	..	..	..	..	..	..	..	..
Others	227	..	..	..	..	..	..	..	..	..
Total	312	27.1	..	..	312	15.4	41	2.3	353	9.2
World Total(f)	1 151	100.0	871	100.0	2 022	100.0	1 827	100.0	3 849	100.0

Sources: Principally OECD-NEA/IAEA, *Uranium: Resources, Production and Demand*, 1975, Tables 1 and 2, amended as follows—

For Canada, G. M. MacNabb, Uranium Canada Limited: *North American uranium resources, policies, prospects and pricing*. Paper given at Uranium Institute, International Symposium on uranium supply and demand, London, June 1976.

For the U.S.A., J. A. Patterson, U.S. ERDA: *Uranium supply developments*. Paper given at Atomic Industrial Forum, Fuel Cycle Conference, Phoenix, Arizona, March 1976.

For Australia, A. J. Grey, Pancontinental Mining Limited: *Australian uranium—will it ever become available*? Paper given at Uranium Institute, International Symposium on uranium supply and demand, London, June 1976.

Notes: (a) Finland, France, West Germany, Italy, Portugal, Spain, Sweden, Turkey, United Kingdom, Yugoslavia.
(b) Includes Namibia.
(c) Includes Central African Republic and Zaire, not shown separately.
(d) Argentina and Brazil.
(e) India, Japan, Korea.
(f) Excludes countries with centrally planned economies.
(g) Includes 108 thousand tonnes as by-product of phosphate and copper production.
h) Includes 300 thousand tonnes for Sweden.

requirements up to 1985. However, there are significant differences between estimates of potential U.S. uranium production in the 1980s.

Table 8 summarises the various estimates submitted to the Commission of uranium production capacity likely to be available over the next decade. Comparison of this information with estimates of uranium requirements shows that the expected uranium production capacity would more than match the requirements contained in the latest ERDA demand estimates for 1985 (as shown in the last column of Table 7). However, it was suggested to the Commission that the U.S. production capacity estimates are too high, and that a figure of 20 000 tonnes of uranium would be more accurate for 1985. Further evidence revealed a variety of views on this matter, suggesting that a great deal will depend on the extent to which U.S. producers consider that the projected demand-supply situation justifies further investment in mining and milling facilities.

The most recent estimates by ERDA which are available to the Commission suggest that stocks in hand at the end of 1975, together with contracts already made for deliveries from U.S. and other sources, will not be sufficient to meet anticipated requirements up to 1985. However, if ERDA's expectations about U.S. uranium production capacity are realised, this shortfall could easily be met by U.S. uranium producers, who could also contribute substantially to stockpiles held by U.S. producers and to the requirements of other countries.

In summary, it is not possible to be certain of the extent to which uranium production capacity in countries other than Australia will be available in the next decade. However, the evidence before the Commission on this subject suggests that sufficient production capacity could be established to meet requirements if Australia withholds supplies during this period, although some doubts have been expressed about the ability of African producers to maintain stable production of uranium. Beyond 1985, substantial increases in uranium production capacity would be necessary to meet currently projected world requirements (excluding the countries with centrally planned economies). The required increase will depend not only on the rate of growth of nuclear capacity, but also on decisions taken on tails assays, reprocessing and the introduction of fast breeder and other advanced reactors.

Longer-term considerations

Considering the position for the remainder of the century, the range of figures for uranium requirements for the year 2000 shown in Table 7, indicates the wide spectrum of possibilities put forward in various projections. Average rates of growth of population and economic activity in the last two decades of the century will be important in determining the uranium requirements. So will the effects of higher energy prices, the results of attempts to conserve energy, changes in the relative costs of nuclear and coal-fired generation of electricity, the development of economically attractive alternative sources of energy, the timing and effectiveness of recycling of uranium and plutonium, and the development of new reactor types. Possible effects of these variables, which have already been discussed in some detail, suggest that the range of possibilities may be even wider than that shown in Table 7.

If commercial fast breeder reactors are introduced, they may eventually have a big influence on uranium demand. Evidence about the likely timing

Table 7

Estimates of Uranium Requirements to 2000 (thousand tonnes uranium)

Requirements	(1) *AUPF*(b)	(2) *Edison Electric Institute*(c)	(3) *AAEC*(d)	(4) *OECD-NEA/ IAEA*(e)	(5) *US ERDA*(f)
United States of America—					
Annual rate—1985	52	35–47	35	..	25–30
—2000	..	79–154	..	..	60–74
Cumulative to 1985	292	209–272	238	..	145–198
to 2000	..	1 087–1 787	..	..	851–1 038
Other(a)—					
Annual rate—1985	68	57–68	56	..	37–43
—2000	..	183–210	..	..	102–123
Cumulative to 1985	380	346–393	332	..	237–278
to 2000	..	2 180–2 525	..	..	1 346–1 600
Total(a)—					
Annual rate—1985	120	92–115	91	82–101	62–73
—2000	..	262–364	140–210	236–313	162–197
Cumulative to 1985	672	555–665	570	513–594	401–476
to 2000	..	3 267–4 312	3 100–3 800	2 974–3 826	2 197–2 638

Column (1) Evidence given to the Inquiry by the Australian Uranium Producers' Forum.
(2) Edison Electric Institute: *Nuclear fuels supply*, 1976.
(3) Evidence given to the Inquiry by the Australian Atomic Energy Commission.
(4) OECD-NEA/IAEA: *Uranium resources, production and demand*, 1976.
(5) Edward J. Hanrahan, U.S. ERDA: *Demand for uranium and separative work.* Paper given at Atomic Industrial Forum, Fuel Cycle Conference, Phoenix, Arizona, March 1976.

(a) excluding countries with centrally planned economies.
(b) 0.275% tails assay 1976–80, 0.3% tails assay after 1981; no recycling.
(c) high estimate 0.3% tails assay, no recycling; low estimate 0.2% tails assay (U.S. only), recycling.
(d) tails assay increasing from 0.2% to 0.3% over 1980–85 period; U recycling from 1979; Pu recycling from 1981 (U.S.) and 1980 (elsewhere).
(e) 0.25% tails assay; high estimate: no Pu recycle, high growth rate of nuclear capacity; low estimate: Pu recycle from 1981, low growth rate of nuclear capacity.
(f) high estimate 0.3% tails assay; low estimate 0.2% tails assay; Pu recycle from 1982 (U.S.) and 1978 (elsewhere); U recycle from 1978; "low" growth scenario.

of their introduction was based mainly on engineering design and development considerations rather than possible economic and political constraints. The most common view was that a commercial fast breeder would not be operating before 1990. However, there were suggestions that the planned French Super-Phenix and British CFR-1 commercial fast breeders might start operating in the late 1980s.

The introduction of fast breeders may be delayed by economic or political factors. The capital cost problems experienced with conventional thermal reactors are likely to be greater with the more capital-intensive fast breeders. Political opposition could stem from, among other things, the fact that the use of breeder reactors will involve the use of highly enriched uranium or plutonium as fuel, as well as their recycling.

The introduction of breeder reactors into electricity generating systems would eventually have a downward impact on demand for uranium. The size of the impact would depend on the doubling time achieved in commercial FBRs. The doubling time is the time required to breed enough nuclear fuel to reload the core of the reactor and supply the core for a second FBR. Evidence suggests that this is unlikely to be less than 15 to 20 years, corresponding to annual growth rates of 4.7 and 3.5 per cent respectively.

If electrical generating capacity grows faster than the doubling rate, it will not be possible to meet the demand for new capacity with a self-supporting program of plutonium-fuelled FBRs alone. Any shortfall in capacity could be supplied by non-nuclear stations or thermal reactors. Alternatively, it would be possible to build FBRs which can use highly enriched uranium fuel instead of plutonium. If either of these two nuclear options is chosen, the demand for uranium will continue to increase, though at a lower rate than if FBRs had not replaced thermal reactors.

One estimate suggested that a ten year delay in introducing fast breeders, from 1990 to 2000, would cause an increase of about 300 000 tonnes in cumulative uranium requirements up to the year 2000, with annual requirements in that year likely to be about 30 per cent above the levels required if there were no delay. The estimate is sensitive to assumptions about future rates of increase in electricity demand, the contribution of nuclear power to meeting the demand, the rate of development and deployment of fast breeder reactors and the reactors' fuel doubling time.

There was disagreement among witnesses about when a significant impact on uranium requirements would be noticed, apparently stemming from disagreement about these assumptions. Some witnesses thought that demand for uranium might start to contract from about the year 2000. Others thought that demand would be largely unaffected until 2010 or 2015 at the earliest. Another judgment was that the market would be secure for hundreds of years; demand might slacken, but would never actually stop.

The weight of evidence indicates that demand for large amounts of uranium for thermal reactors will probably continue well into the next century.

Existing Australian contracts

Contracts for the supply of about 9000 tonnes of uranium as yellowcake have been entered into by various existing and potential Australian suppliers with utilities in Japan, the U.S.A. and the Federal Republic of Germany. These provide for supply of uranium in the years 1976 to 1986 inclusive, and were approved by the Australian Government before December 1972. The present Government has stated that: 'Subject to satisfactory commercial renegotiation of previously approved contracts . . . where necessary, arrangements would be made with Australian companies for access to the Government's uranium stockpile to meet early delivery under those contracts'.

The amounts included in these contracts represent only a small proportion of the projected requirements of the countries mentioned in the next decade. It was suggested by several witnesses that the commitments by Ranger and Queensland Mines up to 1980 could be met by releasing the 1750 tonnes of uranium as yellowcake presently stockpiled at Lucas Heights. Commitments after 1980 could be met by arranging either production by Mary Kathleen or purchase overseas of the balance required, amounting to about 3000 tonnes. Mary Kathleen's present contracts provide for the supply of 4021 tonnes to electricity utilities in the U.S.A., Japan and the Federal Republic of Germany between 1976 and 1982. Alternatively, if it were decided not to export any Australian uranium at all, it seems likely that arrangements could be made for supply by other countries of the full 9000 tonnes, since the amounts required per year are small in relation to total world production.

Table 8

Uranium Production Capacity 1975, 1980, 1985 (thousand tonnes uranium per year)

	AAEC (1)			*OECD-NEA/IAEA* (2)	
Country/Region	*1975*	*1980*	*1985*	*1980*	*1985*
North America—					
Canada	4.6	10.0	12.0–15.0	10.0	11.5
United States	14.0	20.0–25.0	30.0–35.0	25.0	40.0
Mexico	(a)	(a)	(a)	0.3	1.0
Western Europe—					
France	1.7	2.0–3.0	2.0	3.0	3.0–3.5
Other(b)	(a)	(a)	(a)	1.3	3.8–4.3
Africa—					
Gabon	0.8	0.8	1.0	1.2	1.2
Niger	1.2	3.0–4.0	4.0–6.0	4.0	6.0
South Africa(c)	3.5	6.0–8.0	10.0–12.0	11.2	13.8
Other	0.5	2.0	3.0–5.0	0.6	0.7
Total(d)	26.3	43.8–52.8	62.0–76.0	56.6	81.0–82.0

Column (1) Evidence given to the Inquiry by the Australian Atomic Energy Commission.
(2) OECD-NEA/IAEA: *Uranium: resources, production and demand*, 1975.
(a) Not shown separately, included in 'other'.
(b) Including Denmark, Germany, Italy, Portugal, Spain, Sweden and Yugoslavia.
(c) Including Namibia.
(d) Excluding Australia and the countries with centrally planned economies.

Potential markets for Australian uranium

As the country with the largest present and projected installed nuclear generating capacity, the U.S.A. is likely to remain the biggest consumer of uranium. Evaluation of the evidence relating to the American market for uranium is therefore of great importance in attempting to predict likely trends in prices and demand. Information on possible markets for Australian uranium in countries other than the U.S.A. is far less comprehensive.

Opportunities for sales to the United States

Opportunities for Australian producers to contract for the sale of uranium to U.S. buyers for delivery up to 1985 will depend heavily on production capacity in the U.S.A. and on the extent to which U.S. utilities wish to build up stocks. They will also depend on whether the utilities are permitted to, and wish to, arrange further supplies from sources outside the U.S.A. As only part of the estimated requirements for this period have been covered by contracts, it appears that Australian producers could obtain a share of uncommitted requirements up to 1985, particularly for deliveries in later years. This would depend, of course, on their entering the market before other producers contract to supply all such requirements.

The Commission was told that the U.S. Government has expressed a desire to ensure that the country's energy supplies are not subject to interference from external events, and that it is likely to limit the proportion of U.S. uranium requirements met from external sources. The existing total embargo on importing uranium for domestic use is to be gradually relaxed from January 1977, but the Commission was informed that U.S. utilities and government bodies would probably limit their external purchases of uranium to about 20 per cent of total requirements. If this proportion is applied, it seems probable that a maximum of from 5000 to 6000 tonnes of uranium could be imported into the United States each year between 1979 and 1985. The Commission was told that contracts for the import of about 3000 tonnes per year had already been entered into by 1 January 1976. It appears therefore that, at the beginning of 1976, further opportunities to make contracts to export to the United States did not exceed the range of 2000 to 3000 tonnes per year until after 1985. These opportunities will have been further reduced by any contracts entered into by non-U.S. producers since that date to supply American utilities.

Potential markets in other countries

Evidence was given that contracts have already been arranged to meet Japan's total uranium requirements up to 1985, although some utilities have not covered all their requirements while others have arranged to receive supplies that will exceed their needs up to that time. Substantial uncommitted demand appears to exist in Western Europe for delivery in the first half of the 1980s.

The evidence indicates that, by 1985, a very high proportion of annual uranium requirements in countries other than the United States could be met from presently anticipated world production excluding that of the centrally planned economies, the U.S.A. and Australia. The estimated requirements exceed the anticipated production, calculated on this basis, by between 10 000 and 15 000 tonnes per year. To cover this deficit, and meet any demand arising from a desire to increase inventory levels, a higher output may be needed in producing countries.

Considering all the information available to the Commission on this matter, it appears that estimates of potential Australian sales of uranium up to 1985, presented by the Australian Uranium Producers Forum, the Australian Atomic Energy Commission and Pancontinental Mining Ltd, are probably too high. If uranium imports to the United States are limited to a maximum of 20 per cent of its total requirements, the average uncommitted import figure for this period would be about 2000 tonnes per year (see column (6), Table 9). If the whole of this U.S. balance could be filled by Australian suppliers, and Australian producers were able to meet one-third of the total uncommitted demand in Western Europe, total Australian sales could rise from under 4000 tonnes in 1980 to over 11 000 tonnes in 1985 (as shown in column (8), Table 9). Similar sales could be made if no import ceiling were imposed by the U.S. and Australia could supply 25 per cent of all uncommitted demand in the U.S. and Western Europe (column (9), Table 9). If Australia attempted to meet a higher proportion of uncommitted demand, for example half the maximum uncommitted U.S. imports plus half the unfilled Western European requirements, the quantity sold would rise at a somewhat faster rate after 1982 and reach about 15 000 tonnes by 1985. However, since the latter figure would represent about 20 per cent of all estimated uranium requirements by 1985, it seems doubtful whether such a level of exports could be achieved without causing a noticeable decrease in the price of uranium. The Australian Government has stated that it intends to avoid such effects by the establishment of orderly marketing procedures.

After 1985, provided growth of nuclear power continues as currently projected, Australia could be an important supplier of uranium to Japan, Western Europe and the United States. However, the evidence indicates that

Table 9

Potential Uranium Markets, 1980 to 1985 (thousand tonnes uranium)

Year	United States of America						Western Europe	Possible Australian markets		
	(1) Total requirements	(2) Domestic commitments 1 January 1976	(3) Imports contracts 1 January 1976	(4) Uncommitted demand (= (1)–(2)–(3))	(5) Estimated maximum imports (20 per cent of (1))	(6) Estimated uncommitted imports ((5)–(3))	(7) Uncommitted requirements	(8) U.S. uncommitted imports plus one-third Western European uncommitted requirements ((6) + one-third (7))	(9) 25 per cent of all uncommitted U.S. and Western European requirements (25 per cent (4 + 7))	(10) 50 per cent of U.S. uncommitted imports plus 50 per cent Western European uncommitted requirements (50 per cent (6 + 7))
1980 . .	24.2	11.7	3.2	9.3	4.8	1.6	6.1	3.6	3.9	3.9
1981 . .	27.1	9.1	3.2	14.8	5.4	2.2	9.6	5.4	6.1	5.9
1982 . .	27.5	8.1	3.0	16.4	5.5	2.5	14.0	7.2	7.6	8.3
1983 . .	26.2	6.4	3.0	16.8	5.2	2.2	19.3	8.6	9.0	10.8
1984 . .	26.8	4.4	2.8	19.6	5.4	2.6	20.0	9.3	9.9	11.3
1985 . .	24.4	3.3	2.6	18.5	4.9	2.3	27.7	11.5	11.6	15.0

Source: Columns 1 to 3—U.S. ERDA, *Survey of United States Uranium Marketing Activity*, 1976.
Column 7—Evidence given to the Inquiry by J. A. Waddams, Director of Company Planning, British Nuclear Fuels Ltd.

a world shortage of natural uranium is unlikely in the light of existing and probable resources and currently projected demands, so alternative supplies will probably be available to these countries. This conclusion is based on the assumption that there will be sufficient investment in overseas countries to develop additional uranium production capacity. Recent market prices for uranium suggest that such investment is likely to be made.

If currently projected rates of growth in economic activity are not attained, or the competitive position of nuclear power is reduced by expected increases in capital costs of nuclear power stations and in the cost of enrichment and reprocessing, it is possible that long-term demand for uranium could be much lower than currently projected. In this event, if substantial increases in uranium production capacity have occurred, reductions in the real price of uranium will eventuate.

Effects of Australia not supplying other countries

There was general agreement in evidence that if Australia does not supply uranium to the rest of the world in the first half of the 1980s, and if nuclear power programs proceed as currently expected, additional uranium mining and milling facilities will have to be established in other countries. It was argued that, if Australian uranium is made available, its availability may serve to prevent the emergence of higher cost producers in the United States and elsewhere. If supplies from Australia or elsewhere are not made available, such producers may have to come into operation and an increase in the average price for uranium may be needed to make their operations profitable.

The relation between prices of uranium in the U.S. market and those in the rest of the world will depend largely on U.S. import strategies and on the availability in world markets of low-cost uranium. It seems probable that the U.S. will wish to restrict the contribution of imports to its total uranium consumption, so that its own industry will remain viable and a high degree of dependence on imports will be avoided. This suggests that the price of uranium in the U.S. will remain at levels sufficient to allow profitable production from U.S. mines. In this case, and if an excess world supply situation does not emerge, it is probable that uranium producers will seek and obtain prices on the world market close to those prevailing in the U.S. On the other hand, if a substantial excess of supply of uranium develops on the world market, prices outside the U.S. may become much lower than those within that country. It is not possible to predict with any degree of confidence which situation is more likely to prevail. The possibility of major long term variations in future uranium prices must be recognised in the light of historical fluctuations in world prices of uranium and other commodities.

Any attempt to estimate the maximum effect on average prices of withholding Australian uranium must take account of the possibility that this action would result in a raising of the U.S. price to cover the costs of American and other producers using lower-grade ore. However, there is the likelihood that further lower-cost supplies would become available from higher grade ores in other countries to meet any deficiency created by Australia not exporting uranium.

Evidence before the Commission gives the results of a recent study which examined the uranium prices which would probably be required to provide sufficient incentive for U.S. uranium production to increase between 1980 and 2000 to meet current estimates of U.S. uranium requirements.

These estimated prices include allowances for exploration costs and a net after-tax return of 15 per cent on capital invested. The study's estimates of the average prices required are US$18.50 per pound in 1985, US$20 in 1990, and US$40.50 in 2000 (all expressed in 1975 price levels). This study notes that some contracts already entered into provide for estimated prices in 1985 above US$18.50 per pound, but points out that these projected prices include attempts to estimate the effects of future inflation, and their real price in 1975 price levels would be considerably lower.

The evidence suggests that Australian producers, using presently known high grade ores, could supply uranium at prices below these estimates of future U.S. prices. However, Australia's share of the U.S. market seems likely to be constrained by the establishment by the Australian Government of orderly marketing procedures as well as by a U.S. Government limit on imports. Based on current projections of uranium demand, it appears that the entry into the market of Australian producers would be unlikely to cause a large decrease in the price of uranium in the U.S. and, conversely, that withholding Australian supplies of uranium would not appreciably affect U.S. uranium prices, at least until after 1990. The average U.S. price could then rise significantly because of a need to stimulate domestic uranium production. This rise would not occur if demand is much lower than presently predicted, or if new high grade uranium ores are discovered and mined in the United States or elsewhere.

No evidence was presented about prices likely to be required to cover costs in countries other than the U.S.A. The situation is much less certain in many potential producing countries, because the necessary exploratory work has not been done. However, a comparison of currently projected requirements (see Table 7) and low cost resources in countries other than the United States (Table 6) suggests that it may be more difficult for other countries to obtain their uranium requirements at around US$20 per pound of U_3O_8 (in 1976 price levels) in the 1980s without Australian supplies. As in the case of the United States, the unavailability of Australian supplies could cause uranium prices to rise in the 1990s only if the demand for uranium continues to rise and if new high grade ores are not discovered and mined in Africa and elsewhere. In addition, the effects of technical change may be expected to reduce the real costs of uranium mining and milling, and thus make it more unlikely that the absence of Australian supplies of uranium will result in higher real prices.

Overall, the evidence suggests that the effect of withholding Australian supplies on average uranium prices in the rest of the world would be minimal, particularly before 1990. If nuclear power programs proceed as currently anticipated, the principal effect of not making Australian supplies available would be to make it necessary for other countries to put more effort into developing and mining known uranium resources. Beyond that time, the effect on uranium prices would depend to a large extent on developments in the demand for natural uranium and the rate of discovery and development of resources in other countries. Because of the stepped-up exploration and development of uranium resources encouraged by the recent level of prices and the likely influence of technical change, the effects on prices of non-availability of Australian uranium may continue to be minimal after 1990.

Evidence before the Commission suggests that a US$10 per pound increase in the price of U_3O_8 would increase the cost of electricity generation by about 0.1 cent (U.S.) per kilowatt hour. As the price per unit of electricity in the United States in 1975 was approximately 2.7 cents, an increase of 0.1 cent would amount to a price rise of about 4 per cent. It seems unlikely, therefore, that the withholding of Australian supplies of uranium would result in any very significant increase in the price of electricity in other countries. Even if the price of uranium doubled, the effect on electricity prices would not be very great. Such an increase will not occur if further high grade ores are discovered and mined in other countries, as seems likely.

Conclusions

Clearly, the development of nuclear power in the rest of the world can continue whether or not Australian uranium is made available. The evidence indicates that, if nuclear power programs proceed at the rate projected in 1976, additional uranium production capacity will have to be established in other countries to meet projected demands in the 1980s. Sufficient supplies of higher grade ores appear to be available for this purpose. Consequently, the unavailability of Australian supplies in this period would not make an appreciable difference to the average price of uranium. An aggressive marketing policy for Australian uranium, designed to capture a large share of the market, might reduce the price level. But such a policy is contrary to the stated intention of the Australian Government and of the companies which gave evidence to the Commission. On these bases it is concluded that no very significant effect on market price would occur because of the absence or presence of Australian supplies in this period.

Beyond 1990, it may be necessary to explore and develop resources in higher cost categories to meet currently projected requirements of uranium, whether or not Australia enters the market, unless further high grade ores are discovered and mined. But such discoveries seem likely, because of the incentive to new activity provided by the uranium price rises in recent years and the small amount of exploration which has been carried out in some countries with geologically promising areas. If they are made, and technical innovations reduce real costs of uranium production, as seems likely, the absence of Australian supplies may have no substantial effect on average uranium prices up to the year 2000. If, on the other hand, further low cost resources are not developed, the use of lower grade ores will cause increases in the price of uranium. But these seem unlikely to have any very substantial impact on the price of electricity in other countries.

If the increases currently projected in uranium requirements for generating electricity up to the end of this century occur, a world ‘shortage’ of uranium appears possible in the next century. But this may not eventuate if rates of increase in energy use diminish, if alternative energy sources prove more economic, or if the real price of uranium now apparent is maintained or increased and additional discoveries and technical improvements are made.

If Australian uranium is marketed in the rest of the world, it appears that some could be sold in the early years of the next decade without significantly reducing market prices—the quantity increasing from between 2000 and 3000 tonnes in 1980 to between 10 000 and 15 000 tonnes in 1985. Further increases may occur in later years, if nuclear power proves in large areas of the world to be the most economic method of generating base load electricity. Recent analyses suggest that nuclear power may become less competitive in the future, in which case total uranium requirements and markets for Australian uranium would be limited accordingly.

If Australia withholds supplies, the extent of the economic benefits foregone may be illustrated by reference to the net benefits which Australia may derive from uranium exports, considered in the next Chapter. It is noted that major, comparatively long term, variations in 'real' prices for uranium may occur in the future, as they have in the past.

9 BENEFITS AND COSTS OF EXPORTING AND NOT EXPORTING AUSTRALIAN URANIUM

Proponents of mining argued that substantial benefits would accrue to Australia from uranium production and exports. These benefits would include foreign exchange earnings, employment opportunities, taxation revenue, and possible stimulation of Australian industries able to supply goods and services to the mining projects. Many of the benefits would flow directly to the Northern Territory. Opponents of mining argued, on the other hand, that any benefits would be outweighed by environmental losses of various kinds. They put the view that any increase in the ability of beneficiaries to buy goods and services would be offset by reductions in welfare associated with these environmental losses.

It is difficult—perhaps impossible—to attribute monetary values to many of the losses which it was suggested the Commission should take into account. The discussion in this Chapter relates principally to those economic benefits and costs which can be measured in monetary terms. However, this should not be taken as indicating that the Commission regards measurable benefits and costs as necessarily more important than those which cannot be quantified; in the case of the Ranger proposal, the evidence shows clearly that some of the most important possible effects cannot be expressed in monetary terms. This Chapter includes a brief discussion of the difficulties associated with evaluating environmental losses in quantitative terms.

Methodology used in economic analysis

Economic benefit-cost analysis enables a project's potential contribution to national income to be assessed. The methodology normally takes income earned from the sale of output as an indication of gross benefits. It does not generally distinguish between sales made within the country concerned and those made to other countries, because the purpose of exporting goods is basically the same as that of selling to domestic consumers. In the case of exports, the foreign exchange earned enables imported goods and services to be purchased for use in the domestic economy.

The only situation in which special treatment for exports may possibly be regarded as justified is where a country's long run balance of payments prospects appear likely to be persistently unfavourable. Then, it might be appropriate to give potential exports a higher value. Such an adjustment does not seem warranted in Australia's case, because there appears to be no cause for pessimism about Australia's long-run balance of payments prospects. Although fluctuations in receipts and payments of foreign exchange may be expected to occur from time to time, and these may lead to adjustments in the rate of exchange, it does not appear that any systematic depreciation of the Australian currency in relation to other currencies will occur unless the rate of inflation in Australia persistently exceeds that of its major trading partners.

Nevertheless it may be useful to calculate the contributions which uranium exports might make to Australia's potential foreign exchange earnings. When this is done, however, it must be recognised that such con-

24647/76—6

tributions are not additional to the net benefits estimated by benefit-cost procedures.

These procedures may be used to estimate the rate of return on the capital invested in a project, and that rate of return may then be compared with the return likely to be obtained from alternative uses of the capital. By deducting costs incurred from expected gross proceeds, the technique may also be used to estimate the net additions to national income contributed by the project in a particular year. In both cases, the net returns are calculated on a pre-tax basis. The income tax levied by governments on profits is part of the return on the project, not additional to it. So it would be double-counting to add taxation revenue to estimates of net benefits accruing to owners of capital resident in Australia. Where foreign-owned capital is employed, on the other hand, taxation revenue is generally the main source of gain to the home country, so separate estimates of economic benefits accruing to Australia are necessary when foreign ownership is important.

Benefit-cost analysis normally attributes a cost to all inputs of resources which could otherwise have been used elsewhere in the economy; the accounts of private or public enterprise treat such costs in a similar manner. Although adjustments need to be made in some cases to allow for differences between costs actually incurred by businesses and the national cost of inputs, the deduction of input costs from gross benefits reflects the opportunities lost in not employing the labour, capital and other inputs in alternative uses.

It may be useful to supplement the results of calculations of this type with other indicators of the economic implications of projects. For example, information on the possible effects of a project on employment opportunities may be required, particularly if the creation of employment opportunities comes to assume a more important role in government policies towards particular industries than it has done in recent decades. Since the Second World War, the Australian Government has relied mainly on its fiscal and monetary powers to deal with general employment problems in Australia. The total employment created if the uranium mining projects go ahead will be small in relation to the size of the Australian labour force. The main impact would, of course, be in the Northern Territory; employment and other regional economic aspects of the mining proposals will be discussed in more detail in the Commission's Second Report.

So-called 'external' benefits and costs are not included in many benefit-cost studies. These 'external' effects may be generally thought of as benefits and costs which affect well-being but which do not show up in financial transactions. As discussed below, they may be extremely important in the case of uranium mining.

Deficiencies in benefit-cost approach

The traditional view, embodied in most benefit-cost studies, is that land not currently used for the production of commodities does not represent a valuable resource; its development, therefore, is assumed not to incur an opportunity cost in the form of losses suffered by those who value it in its undeveloped state.

Evidence before the Commission indicates that methodologies are being developed which attempt to take more adequate account of the consequences of development proposals in areas containing valuable or potentially valuable environmental attributes.

The use of areas for uranium mining and associated activities, both in the more immediate future and over a very long span of time, may involve important consequences. While uncertainty exists about man's ability to control harmful effects of the use of uranium, account must be taken of losses which may be suffered if people are harmed directly or indirectly by the mining and use of uranium. Similar considerations apply to any losses incurred by aborigines or others to whom the area affected by mining has a special significance.

Concern was expressed to the Commission about possible harmful effects from mining on the proposed Kakadu National Park. It is relevant, in considering these possible effects, that the growth in real incomes per head in developed countries in recent decades has increased the demand for services provided by the natural environment. There has been a rapid growth in demand for the use of areas such as national parks where people can make closer contact with the natural environment.

While a substantial amount of evidence dealt with losses expected to result from a decision to allow the production and export of uranium, evidence also pointed to some favourable 'external' effects. For example, facilities provided by the mining companies or associated with the proposed regional centre might be used by members of the public to gain access to the area for recreational purposes. Attaching values to the benefits provided by such facilities is also very difficult. Future uses of the facilities would have to be considered in some detail and studies made of the potential value of the facilities over time.

To summarise, the evidence points to many aspects of the Ranger proposal that the present state of the art of benefit-cost analysis does not allow to be adequately quantified. In particular, the proposal, and other proposed developments such as additional uranium mines and mills and the regional centre, would have very significant effects on the aborigines and on the present 'natural' character of the region. These could include both benefits and losses, but the evidence suggests that the losses would predominate. Other possible losses include ecological effects such as changes in species diversity and distribution, and a possibly irreversible commitment to further development, triggered by the initial development, which would reduce the choices available to future generations. Possible advantages include an increased opportunity for people to visit the area for recreational, education or other purposes. Associated with these would be the disadvantages of damage to aboriginal sacred sites, pollution, litter, damage to fauna and flora, and erosion. These will be considered in the Commission's Second Report.

Measurable benefits and costs of the Ranger proposal

The data in Table 10 illustrate the application of benefit-cost analysis to the potential operation of the first stage of the Ranger project up to 1989-90, involving the annual production of 3000 tonnes of U_3O_8. The table is based on financial estimates provided to the Commission by the proponents, and indicates the joint venturers' expectations of financial results. January 1976 price levels are used throughout. Any inflationary effects occurring after January 1976 which cause prices and costs to rise at the same rate will not affect the real return from the project but will merely change its monetary value in accordance with changing price levels.

It was assumed, in preparing the estimates, that the main capital expenditure would be incurred from 1976-77 to 1979-80, that the plant would be

Table 10

Estimates of Revenue and Expenditure from Ranger Project, 3 000 tonnes U_3O_8 per annum, to 1989–90
($A million; in January 1976 prices)

Year	(1) *Revenue*	(2) *Capital Expenditure*	(3) *Operating Expenditure*	(4) *Revenue— Expenditure* (1)–(2)–(3)	(5) *Present worth factor* (10 *per cent Per annum*)	(6) *Present worth* (4) *x* (5)	(7) *Cumulative present worth*
1976–77 . .	0	13.5	0	−13.5	1.000	−13.5	−13.5
1977–78 . .	0	32.2	0	−32.2	0.909	−29.3	−42.8
1978–79 . .	0	49.7	0	−49.7	0.826	−41.1	−83.9
1979–80 . .	9.4	22.3	12.8	−25.7	0.751	−19.3	−103.2
1980–81 . .	58.7	2.4	26.5	29.8	0.683	20.4	−82.8
1981–82 . .	97.2	0.7	27.3	69.2	0.621	43.0	−39.8
1982–83 . .	98.4	1.5	27.3	69.6	0.564	39.3	−0.5
1983–84 . .	99.6	3.0	27.3	69.3	0.513	35.6	35.1
1984–85 . .	99.8	2.2	27.3	70.3	0.466	32.8	67.9
1985–86 . .	102.4	5.8	27.4	69.2	0.424	29.3	97.2
1986–87 . .	104.8	4.7	27.4	72.7	0.385	28.0	125.2
1987–88 . .	104.8	1.0	27.4	76.4	0.350	26.7	151.9
1988–89 . .	104.8	2.4	27.4	75.0	0.319	23.9	175.8
1989–90 . .	104.8	3.6	27.4	73.8	0.290	21.4	197.2
Totals .	984.7	145.0	285.5	554.2	..	197.2	..

completed by October 1979 and that production would commence shortly afterwards.

The revenue figures shown in column (1) of Table 10 assume a constant selling price of US$20 per pound of U_3O_8 in yellowcake, with conversions to Australian currency at the rate of US$1.26 = $A1.00. They also assume that the plant operates at capacity and that all output is sold in the year in which it is produced; no costs are included which would be associated with slowdowns in production or with any stockpiling of yellowcake due to sales falling short of production. To the extent that these assumptions are not realised in practice, the net benefits are over-estimated.

Estimates of capital expenditure include costs of all plant, equipment, buildings and services at the site and a contribution to costs of infrastructure such as housing. These estimates are based on engineering studies carried out in 1972-73, and are updated to January 1976 price levels by the use of price indexes. The proponents indicated that presently unknown factors encountered during construction may raise capital expenditure above the estimated level.

Operating cost estimates were also derived by updating data prepared in 1972-73, and are subject to the same qualification with regard to accuracy. The expenditure figures include estimates of charges by government authorities to cover costs of services, to the extent that these costs would normally be recovered from users. Additional costs which may be incurred with the establishment and operation of the proposed regional centre, and any environmental or other costs not envisaged in the Environmental Impact Statement, are not included. The estimates in Table 10 therefore understate national costs by the amount of any unrecovered government expenditure. Similarly, the costs to others of preventing environmental damage and any other losses incurred as a result of the project proceeding would have to be deducted from the estimates presented in the table. The Commission's Second Report will consider these aspects in more detail.

Working from the data in Table 10, the proponents presented calculations showing a 32 per cent rate of return on capital before income tax, indicating a high level of profitability for the project if the average real price of U_3O_8 assumed—US$20 per pound in January 1976 price levels—is achieved. A 25 per cent fall in the average real price reduces the rate of return to 22 per cent, and a 25 per cent increase in real price increases it to 39 per cent.

The figures in column (4) of Table 10 show the difference between anticipated revenues and costs, without allowing for interest on capital. Profits will be reduced to the extent that interest is paid on capital outlays incurred before any revenue is received. They will also be reduced to the extent that expenditure was incurred on exploration and other activities before 1976-77. On the assumptions embodied in the data, the project would show a positive cash flow (the difference between revenue and expenditure in a given time period) in 1980-81. Payment of income tax would delay the receipt of a net cash flow to the companies concerned until 1982-83. Under the terms of the Memorandum of Understanding, the Commonwealth Government would incur a maximum cash deficit of about $85 million (in January 1976 prices) in 1980. It would receive a positive net cash flow commencing in 1982-83, and accumulate a substantial net return from its interest in the latter half of the 1980s.

Column (6) of Table 10 shows calculations of the present worth of the net profits derived from these data. The calculations show that the additional net income generated, when discounted at a rate of 10 per cent per year to the financial year 1976-77, is approximately $197 million. The cost of resources used in the construction stage is effectively recovered by 1982-83, if the assumptions involved in the calculations eventuate. After that time, the project would make a positive net contribution to national income.

To put this possible contribution in perspective, the present worth of net additions to national income generated by the project can be compared with the present level of national income. In 1974-75, the most recent year for which figures are available, Australian national income was approximately $54 500 million. At January 1976 price levels, this is approximately $58 000 million. So the present worth of total additions to national income from the first stage of the Ranger project, based on the above calculations, would be about 0.34 per cent of recent annual levels of Australian national income.

The Ranger operation would employ about 600 people during the construction period of two years, if the initial production rate were 3000 tonnes per year of U_3O_8 in yellowcake and 1000 people if the rate were 6000 tonnes per year. Thereafter the operation would employ about 250 and 400 people respectively. The evidence indicates that the great majority of those employed in the construction phase would be skilled or semi-skilled workmen from the south, and that the permanent workforce at the mines would also have a preponderance, although a lesser one, of such people. Opportunities would exist for employment of limited numbers of people from the Northern Territory, both Aboriginal and European, mainly after completion of the construction phase.

Although there is a possibility that world prices for uranium, and consequently the profitability of the venture, may decline, it is concluded,

Table 11

Estimates of Possible Increases in Exports and Net National Income from Uranium Production, 1980-81 to 1999-2000

(All monetary figures in January 1976 prices)

Year	(1) *Quantity (thousand tonnes uranium)*	(2) *Revenue*	(3) *Costs*	(4) *Estimated profits (before tax) (2) – (3)*	(5) *Estimated increase in net national income 87.5 per cent of (4)*	(6) *Estimated level of national income*	(7) *Estimated percentage contribution to net national income* $\frac{(5)}{(6)} \times 100$	(8) *Estimated value of exports of goods and services*	(9) *Estimated percentage contribution to total export income* $\frac{(2)}{(8)} \times 100$
		$ million	$ million	$ million	$ million	$ thousand million		$ thousand million	
1980–81 . . .	2.5	100.0	45.0	55.0	48.1	70.6	0.068	14.0	0.71
1981–82 . . .	5.0	200.0	90.0	110.0	96.3	73.4	0.131	14.7	1.36
1982–83 . . .	7.5	300.0	135.0	165.0	144.4	76.4	0.189	15.4	1.95
1983–84 . . .	10.0	400.0	180.0	220.0	192.5	79.4	0.242	16.2	2.47
1984–85 . . .	12.5	500.0	225.0	275.0	240.6	82.6	0.291	17.1	2.92
1985–86 . . .	15.0	600.0	270.0	330.0	288.8	85.9	0.336	17.9	3.35
1986–87 . . .	17.5	700.0	315.0	385.0	336.9	89.3	0.377	18.8	3.72
1987–88 . . .	20.0	800.0	360.0	440.0	385.0	92.9	0.414	19.7	4.06
1988–89 . . .	22.5	900 0	405.0	495.0	433.1	96.6	0.448	20.7	4.35
1989–90 . . .	25.0	1 000.0	450.0	550.0	481.3	100.5	0.479	21.7	4.61
1990–91 . . .	27.5	1 100.0	495.0	605.0	529.4	104.5	0.507	22.8	4.82
1991–92 . . .	30.0	1 200.0	540.0	660.0	577.5	108.7	0.531	23.9	5.02
1992–93 . . .	30.0	1 200.0	540.0	660.0	577.5	113.0	0.511	25.1	4.78
1993–94 . . .	30.0	1 200.0	540.0	660.0	577.5	117.6	0.491	26.4	4.55
1994–95 . . .	30.0	1 200.0	540.0	660.0	577.5	122.3	0.472	27.7	4.33
1995–96 . . .	30.0	1 200.0	540.0	660.0	577.5	127.1	0.454	29.1	4.12
1996–97 . . .	30.0	1 200.0	540.0	660.0	577.5	132.2	0.437	30.6	3.92
1997–98 . . .	30.0	1 200.0	540.0	660.0	577.5	137.5	0.420	32.1	3.74
1998–99 . . .	30.0	1 200.0	540.0	660.0	577.5	143.0	0.404	33.7	3.56
1999–2000 . .	30.0	1 200.0	540.0	660.0	577.5	148.7	0.388	35.4	3.39

on the basis of data submitted to the Commission, that the Ranger proposal would probably be a highly profitable venture. It would probably generate an increase in net national income which, although substantial in relation to the investment directly required, would be small in relation to total national income. Its contribution to employment in Australia would be small compared to the total Australian workforce. It is possible that it and associated developments would have relatively important economic, and other, effects on the region. These will be discussed in the Commission's Second Report.

Application to expansion of Australian sales of uranium

Little evidence was presented to the Commission relating to the costs which would be incurred in increasing the output from the Ranger project or from other uranium ventures. Despite this, it seems desirable to make some estimates of the possible net economic benefits of such increases and to assess the significance of these benefits in relation to future national income. This procedure helps to throw light on the extent to which the development of a substantial Australian uranium producing and exporting industry will contribute to national income. It also shows the size of the measurable economic benefits which may need to be weighed against total environmental considerations.

Table 11 contains data based on possible levels of Australian exports of uranium from 1980-81 to 1999-2000, together with the assumptions, discussed earlier, about costs derived by extrapolating information provided by Ranger. Column (1) shows the quantities of uranium which Australia may be able to export at an average price of about US$20 per pound of U_3O_8 (in January 1976 price levels). The figures for the first six years are consistent with the discussion in Chapter 8. From the mid-1980s, further expansion of exports would appear to be possible without depressing real market prices if total requirements, including those of Japan, expand as suggested by current projections. The data assume an expansion of 2500 tonnes in Australian production and sales each year, reaching a maximum level of 30 000 tonnes per year by 1991-92. This appears to represent a reasonably realistic upper limit to annual Australian production, based on present knowledge of high grade ores, since a continuation of production at this level in the 1990s would exhaust the currently known Australian resources (as shown in Table 6) by about 1997-98. However, as further high grade discoveries seem likely to be made, the data in the table are continued up to the end of the century.

In calculating the total revenue which may be received, an average price of $A40 000 per tonne of uranium was used. This is equivalent to about $A18.14 per pound of uranium, or about $A15.37 per pound of U_3O_8, which in turn is approximately equal to US$20 per pound of U_3O_8 at recent rates of exchange. Revenue actually received will vary with fluctuations in average prices and rates of exchange between Australian and other currencies. As mentioned in Chapter 8, a number of factors suggest that average prices may fall below this figure, so the estimates in this section may be regarded as the maximum returns likely to be achieved. It is important to note also that the rate of sales being assumed is one which it is estimated would not force the price down below levels which would otherwise be necessary to support U.S. producers. Higher levels of Australian sales could probably be made in earlier years, but only at the cost of depressing the real market price, resulting in lower average prices for all

Australian production of uranium. The possibility exists that substantial falls in world uranium prices may occur, regardless of the rate of Australian export.

The cost levels assumed in column (3) of Table 11 are based on the estimates, mentioned earlier, made by the Ranger partners. To enable computations to be made on the basis of annual data, capital costs were assumed to be spread over the estimated years of production. To allow for the opportunity cost of capital required, interest at the rate of 10 per cent per year payable on accumulated cash flow deficits was assumed. This procedure gives a total cost figure, including operating costs, of about $A18 000 per tonne of uranium, which is equivalent to about $A6.9 per pound of U_3O_8 in yellowcake.

This average cost figure (in January 1976 price levels) was assumed to apply to all levels of potential output shown in Table 11. Although there was some evidence that economies of scale would flow from higher levels of production from the Ranger and Pancontinental deposits, no details were provided to the Commission. Neither was any detailed evidence given about costs relating to other producers. It therefore seems likely that costs may be overstated and profits understated in Table 11 to the extent that average costs would be lower with larger scale production. On the other hand, average costs will be higher to the extent that the assumptions made that all mining and milling operations would produce uranium continuously, and that the product would be sold in the year of production, are not met in practice.

The calculation of pre-tax profits in column (4) of Table 11 follows directly from the assumptions of revenues and costs already outlined. However, in considering the possible expansion of Australian production as assumed in the table, account must be taken of the effect of overseas ownership of profits on contributions to net national income. In the case of foreign ownership, these contributions would be derived mainly from income taxes and other levies on foreign-owned profits. Consequently, both the extent of foreign ownership of profits and the extent of likely gains arising from taxes on such profits need to be considered.

The evidence suggests that less than 10 per cent of the equity shares in E.Z. Industries and Peko-Wallsend are owned directly by non-residents, although there is doubt about the ownership of large holdings in the names of nominee companies. Taking into account the Australian Government's potential interest in the Ranger venture, it seems reasonable to assume that overall foreign ownership of Ranger profits would be small.

In the case of Pancontinental, the evidence suggests that overseas ownership amounts to about 49 per cent of equity holdings. It follows that, if Ranger and Pancontinental were both operating, a significant level of foreign ownership of profits would need to be taken into consideration in estimating net national benefits. Moreover, it appears that substantial overseas ownership may be involved in other uranium mining ventures.

The proportion of foreign ownership of profits applying at any time will depend, of course, on the ownership of the companies concerned, but, for the purposes of making the calculations in Table 11, a 25 per cent level was assumed to apply throughout the period. The estimated increases in net national income in column (5) are equivalent to 87.5 per cent of the estimated profit figures in column (4), it being assumed that half the 25 per cent of foreign-owned profits will contribute to national income, by way of income and withholding taxes, royalties and other government levies. This assumption is reasonably consistent with current taxation rates.

As shown in the Table, the estimated contributions to national income rise each year up to the early 1990s. The Table's column (6) shows estimates, commencing in 1975-76 and expressed in January 1976 prices, of Australian national income likely to be realised if a constant 4 per cent per year compound real growth rate is achieved. This assumed growth rate is slightly lower than the average achieved in the last two decades; it makes some allowance for the slower rate of population growth which seems likely to occur in future.

The actual level of national income achieved would obviously depend on many factors which cannot be canvassed in this report, but the figures given in the table may be useful in placing potential economic gains from uranium mining in perspective. As shown in column (7), on the assumptions made in Table 11, potential additions to national income would rise from less than 0.1 per cent of projected national income in 1980-81 to about 0.5 per cent in 1990-91. Thereafter, with production, revenue and costs assumed constant in real terms, the percentage contribution from uranium would fall to about 0.4 per cent by the end of the century. These figures also provide an estimate of the income which would be foregone if Australia does not export any uranium during the period.

The data in Table 11 also make it possible to assess the significance of the possible contribution of uranium exports to Australia's foreign exchange earnings. Column (8) contains estimates of total earnings from all exports of goods and services, based on a 5 per cent per year compound growth rate commencing in 1975-76 from a level of about $11 000 million. This growth rate is considerably less than that achieved by Australia in the last decade, but may be consistent with possible slower growth in the world economy in the rest of the century. As shown in column (9), estimated export proceeds from uranium, based on the assumptions made, would grow from less than one per cent of the projected total in 1980-81 to 5 per cent in 1991-92. They would decline to about 3.4 per cent at the end of the century. A decision not to export uranium would reduce Australia's ability to purchase imports in the same proportion. However, as pointed out previously, the value of possible exports is included in the estimated effect on net national income and is not additional to it.

Effects of increasing export of coal for electricity generation

It was submitted in evidence that continued reliance on fossil fuels for generating electricity, combined with a major program of energy conservation and development of renewable energy sources, could provide a bridge to the time when a high proportion of all energy requirements will be derived from renewable resources. It was also suggested that Australia could help meet the requirements of other countries for fuel to generate electricity by making available some of its extensive reserves of coal suitable for this purpose, rather than by mining and exporting uranium. A number of witnesses suggested that at least some of the economic benefits which Australia might forego by not exporting uranium could be recouped in this way.

There is no doubt that Australia possesses very large reserves of suitable coal, and these are likely to exceed its own requirements for a very long time to come. An outstanding advantage of these coal reserves—particularly the very substantial quantities of black coal known to exist in Queensland and

New South Wales—is that they have a low sulphur content. They are therefore potentially very important from an environmental point of view.

Australian exports of coking coal are at present a major source of export earnings. Exports of coal suitable for electricity generation have generally been on a much smaller scale, and have filled short-term gaps in overseas requirements rather than meeting long-term contracts. In its submission to the Inquiry, the Joint Coal Board indicated that the Australian coal industry is capable of building up a substantial export trade in steaming coal suitable for electricity generation, provided it can obtain markets. It also indicated that such a development would yield substantial economic advantages to Australia, but added that it would need to occur on a long-term basis, so that electricity authorities overseas could ensure that the generating capacity installed made the most efficient use of the coal available.

It is clear that the development of large scale exports of steaming coal would require heavy investment in mining, transport and other facilities. Also it is clear that long-term contracts would be the most appropriate means of ensuring that the necessary facilities were made available in Australia and in the recipient country, and that the necessary shipping capacity was provided. As the evidence suggests that no rapid increases in generating capacity will be needed in the short-term, time appears to be available to investigate thoroughly the possibility of substantially increasing Australian exports of coal for this purpose. It is understood that the Australian Government has already made it known to some potential importers, including Japan, that Australia is interested in developing these coal resources and making them available on a long-term contract basis. Japan is understood to be interested in substantially increasing its imports of this type of coal in the next decade.

Exports of steaming coal by Australia could minimise any fuel shortage which would otherwise arise if other countries decide to reduce their reliance on nuclear power below present expectations. As mentioned in Chapter 7, some increases in electricity prices might follow the substitution of domestic coal for nuclear power in many parts of the world. Such price rises would be lower if the landed cost of Australian supplies proved to be lower than the cost of local coal. This may be particularly important in the case of Japan, in view of the relatively small amounts of coal available from its own resources. However, the cost of transporting coal to Japan would probably be a major factor in determining the outcome.

No evidence was presented to the Commission on the extent of the benefits and costs which would accrue from the production and export of coal for electricity generation purposes. Some witnesses expressed the view that such exports would not be competitive with exports of uranium. They based this conclusion on the argument that the margin of economic advantage held by nuclear power in the principal consuming countries is so large that the provision of Australian coal to these countries would not make coal-fired generating stations economically competitive. However, as discussed in Chapter 7, the margin of costs between these two alternatives is not large in many localities, and any economic advantage held by nuclear power could possibly be reversed if suitable coal were available from Australia.

Conclusions

The evidence before the Commission shows that, provided nuclear power programs proceed as recently projected and uranium prices do not fall considerably below their recent levels in real terms, the Ranger project would probably generate a substantial rate of return on the capital invested. However, its contribution to net national income and employment opportunities would be relatively small.

Assuming the continuance of large nuclear power programs in the economically advanced countries and the marketing of Australian uranium in such a way as to minimise the effect on average uranium prices, net additions to Australian national income arising from a gradual expansion of sales up to an annual rate of 30 000 tonnes of uranium would probably result in an increase of about 0.1 per cent of national income in 1980-81, rising to about 0.5 per cent in 1990-91. The contribution would decline to about 0.4 per cent of national income at the end of the century if the rate of uranium production and exports was maintained and the real price of uranium remained unchanged. On these assumptions, foreign exchange earned by uranium exports would probably contribute about 1 per cent of total export earnings in 1980-81, rising to a maximum of about 5 per cent in 1991-92 and declining to about 3.4 per cent by the end of the century.

There is insufficient evidence from which to assess the likelihood of major decreases in the real world price of uranium in the future. However, further reductions in rates of growth of nuclear power programs could occur, reflecting the less favourable economics of nuclear power implied by recent economic analyses of the nuclear fuel cycle, and lower economic growth rates throughout the world. The present level of uranium prices will probably result in the discovery of large additional high grade uranium deposits in various countries. If these are developed and demand for uranium does not increase at the rates projected, there could be an over-supply situation leading to substantial reductions in world uranium prices. In that case, net additions to Australian national income would be lower than those estimated.

Plate 3. A U.S. gaseous diffusion enrichment plant at Portsmouth, Ohio. One of three large enrichment facilities in the U.S., the plant occupies about 90 hectares and has a demand for electricity in excess of 1,800 megawatts. (Photo by courtesy of the U.S. Energy Research and Development Administration.)

Plate 4. A cylinder of uranium hexafluoride, enriched in uranium-235, is loaded at the Oak Ridge gaseous diffusion enrichment plant, Tennessee, U.S.A., for transport to a fuel fabrication plant. (Photo by courtesy of the U.S. ERDA.)

10 HAZARDS OF THE NUCLEAR FUEL CYCLE

This chapter deals with hazards associated with the normal operation of the fuel cycle and the possible effects of accidents. Risks arising from acts of terrorism and the risks of nuclear war arising from the use by governments of nuclear materials are dealt with in Chapters 12, 13, 14 and 15.

We begin with a general discussion of radiation, which is present to some degree at all stages of the fuel cycle. Then follows a discussion of hazards arising at each stage of the nuclear fuel cycle.

Radiation Radiation from nuclear material, comprising alpha, beta, gamma and neutron components (refer to Chapter 3 for definitions), is of the same form as that which pervades the biosphere and to which mankind has always been exposed. The average background radiation dose received by each person on the surface of the earth is about a tenth of a rem per year. Radiation levels vary from place to place with altitude, and with differences in the amounts of radioactive material in rock and soil, so that in some places the radiation level is as high as thirty times that in other areas.

People may also be exposed to radiation arising from man-made sources, such as medical radiology, certain industrial processes, fallout from nuclear bomb tests, various consumer items including luminous watches and television sets, and nuclear power production. The average doses received by the population from these sources vary from country to country. Reliable figures for Australia are not available, but estimates for the United Kingdom and the U.S.A. are 0.04 and 0.08 rem per person per year respectively. Medical radiology is by far the biggest contributor.

The cells of the human body can be affected by radiation reaching them either from sources outside the body or from sources which have entered the body, principally through the mouth or nose. The results of radiation exposure, whether affecting the people exposed (somatic) or their descendants (genetic), have been widely documented, but the processes involved are complex and not yet fully understood. Differences in interpretation arise when information obtained from high doses and high dose-rates is extrapolated to low doses and low dose-rates.

The length of time between radiation exposure and the appearance of biological symptoms varies. Large doses of radiation (greater than 100-200 rem) delivered rapidy to a large part of the body injure or kill a substantial number of cells, and lead within days to obvious injury or death. The delayed effect is principally the induction of various forms of cancer. This may occur in survivors of large doses of radiation, and it may possibly arise from low doses of radiation which are too small to cause early symptoms. The time between exposure to radiation and the appearance of leukaemia caused by it averages about eight years, while the delay for other types of cancer is generally longer. The available information indicates that, in general, the probability of developing cancer is linearly related to the radiation dose received. It is estimated that, if a population of 10 000 is exposed to whole-body doses of 100 rem, 15 or 20 cases of leukaemia and perhaps 100 cases of all forms of cancer will result within the first 25 years.

The genetic effects of radiation are produced by the mutation of genes in the reproductive cells. Such mutations can also be produced by agents other than radiation. The consequences of such mutations will show up, if at all, one or more generations later. On the basis of limited evidence, geneticists estimate that the increased radiation dose to the gonads (sexual organs) of the whole population not yet past child-bearing age which would double the present rate of mutations lies between 20 and 200 rem, depending on the rate of exposure.

Knowledge of the effects of radiation in causing disease is based on observations of the effects of relatively high doses, such as those received by Japanese atom bomb survivors. Several witnesses suggested that the linear extrapolation of these effects downwards to low radiation doses sets an upper limit to the possible effects of these low doses, and in almost all cases would considerably overestimate the effects. However, some evidence indicates that use of linear extrapolation could, on the contrary, underestimate the effects of alpha radiation at low dose rates.

Biological recovery appears to reduce the risk of delayed injury when a dose is spread over decades as compared with the risk from the same dose received over a short time.

Many witnesses pointed out that the consequences of low radiation doses cannot be observed experimentally, because the effects, such as, for example, an increase in the number of people contracting leukaemia, are so small that they could have arisen through chance alone. This may well explain why, with one possible exception, there has been no evidence of variations in human health, which could be attributed solely to radiation, between regions of the world with particularly high background radiation and regions with a normal or low background. The exception has been recorded recently in India (*Nature 262*, 60-61, July 1976), where observations were made on communities in Kerala which were exposed to background radiation of 1.5 to 3 rem per year. By comparison with people living in the parts of Kerala with low background, the exposed population had an increased frequency of genetic disorders.

Radiation protection standards

It is assumed in setting radiation protection standards that any exposure to radiation involves some risk of injury (i.e. there is no threshold dose), and that the risk is proportional to dose and independent of dose-rate. On the basis of cost-benefit analysis, the International Commission on Radiological Protection (ICRP) recommends radiation protection standards at a sufficiently low level that any further reduction in risk would not be considered by the ICRP to justify the effort required to accomplish it.

The limits set for workers in an industry associated with radiation, and expressed in terms of the amount of radiation that an individual may receive in his occupation during the course of a year, are shown below.

Table 12
Annual Maximum Permissible Doses for a Radiation Worker

Organ or tissue	*Annual maximum permissible dose (rem)*
Gonads, red bone-marrow, whole body	5
Skin, bone, thyroid	30
Hands and forearms, feet and ankles	75
Any other single organ	15

The ICRP recommends that the dose limits for individual members of the public should be set at one-tenth of the corresponding values for radiation workers. It also suggests that, to minimise any long-term genetic effects, the average dose received by members of a large population group should not exceed 0.17 rem per year. Working from those recommendations, the ICRP derives permissible limits for body burden and intake of radioactive material by ingestion or inhalation. It excludes from these limits doses received from natural background and from medical procedures. But it recommends that any unnecessary exposure be avoided and that all doses be kept as low as readily achievable, economic and social considerations being taken into account.

The recommendations of the ICRP are incorporated in the legislation of the States of Australia (but not the Northern Territory or the A.C.T.), and in the *Code of Practice on Radiation Protection in the Mining and Milling of Radioactive Ores* (Australian Department of Health, 1975), which has at this time no legal status. The Commission regards the code as an authoritative document which should be incorporated in legislation.

The ICRP reviews its recommendations as new radiobiological data become available. The existing limits have remained unchanged for a decade. A comprehensive review is currently being undertaken, but in 1972 the ICRP indicated that information then available led it to re-affirm its previously recommended dose limits.

All the dose limits mentioned above apply to situations other than accidents. The ICRP recognises that industries need guidance in setting radiation limits applicable to accident situations. The intention is that, if these are reached as a result of accidents, special action will be taken to protect the people exposed. Various institutions have set limits of this type for themselves.

Radon This gas, produced by the radioactive decay of radium and released in the mining and milling of uranium, has a half-life of 3.8 days. Experience overseas has shown that, in the underground mining of uranium, the lung is the part of the body receiving the highest proportion of allowable dose and is therefore the critical organ. The higher-than-normal incidence of respiratory cancer that has been recorded in some parts of the world among underground miners is attributed to the continued inhalation of air containing high concentrations of the short-lived alpha-emitting decay products of radon.

Precise calculation of the radiation dose received by lung tissues when these decay products are inhaled is difficult. To cope with the problem, radiation protection standards based on the relation between the estimated intake of the decay products and the incidence of lung cancer amongst underground miners have been adopted by some countries, including the U.S.A., but not yet by Australia. These standards are based on the intake during a miner's assumed thirty-year working life-time, and are expressed in terms of Working Level Months (WLM). One WLM is the dose received from breathing air containing decay products at a concentration of 1 Working Level throughout the working portion of a month (taken at 170 hours). A Working Level (WL) is the quantity of radon decay products which, in one litre of air, would deliver a specified total amount of alpha particle energy (equivalent to a ten-thousand millionth of a curie of radon decay products in a situation where radon and the decay products are in equilibrium).

The Australian *Code of Practice* recommends a maximum radon decay-product intake of 4 WLM per year, corresponding to an average concentration of 0.33 WL throughout the year. This standard is as restrictive as that in force anywhere in the world, and corresponds to a cumulative intake of 120 WLM over a thirty year period.

In a study of uranium miners in Colorado, the U.S. National Institute for Occupational Safety and Health has attempted to work out the relation between lung cancer incidence and radon exposure. In miners exposed to a total dose of 120 WLM or less, no increase over the normal lung cancer incidence has been detected. Miners exposed to higher doses have shown a higher than normal incidence.

The study indicates that the risk of contracting lung cancer is about eight times greater for miners who smoke cigarettes than for non-smoking miners. Calculations based on linear extrapolation put the risk of lung cancer to the average smoking miner exposed to the 4 WLM per year limit at about 4 in 10 000 per year above that to equivalent smokers not exposed to radon decay products. This was stated to be the same risk as that arising from smoking an additional one or two cigarettes a day. The Australian *Code of Practice* states that 'Employees should not smoke in underground mines', but it was acknowledged during the Inquiry that it would be difficult to persuade miners not to smoke.

The Commission notes that the suitability of the recommended 4 WLM per year standard for all miners was not disputed during the Hearings.

Plutonium

Plutonium is a man-made element. Virtually none exists in nature, but it is produced from uranium-238 in the normal operation of power reactors. Many witnesses expressed concern about the possible hazards resulting from its potential use as a nuclear explosive and its extreme radiotoxicity.

Plutonium-239 is the isotope of greatest concern; it is the type used in atom bombs and constitutes at least 70 per cent of the total amount of plutonium produced in power reactors. It has a half life of 24 400 years. Much evidence points to the relatively great quantities of plutonium which would be generated and processed in a major world nuclear power program, particularly if fast breeder reactors are widely adopted. At present, about 20 tonnes are produced each year, most of which remains in unprocessed form in spent fuel rods. By the year 2000, the annual production could be several hundred tonnes. That is probably about the quantity of separated plutonium in storage around the world today in the form of nuclear bombs.

Only about 6 kilograms of plutonium-239 are needed in metal form to produce a fission explosion, perhaps 9 kilograms as plutonium oxide, and slightly larger amounts of reactor grade plutonium containing various plutonium isotopes. The danger exists that if sufficient plutonium of reactor grade came together inadvertently, possibly during reprocessing or fuel fabrication, a chain reaction could occur with consequent emission of a powerful pulse of lethal radiation and dispersion, possibly violently, of the plutonium. Strict control measures are routinely applied and have so far proved effective.

As a poison, plutonium in small amounts acts by causing cancer. In soluble form it can be taken into the body through the digestive tract or through a cut, and may produce cancer in the tissue where it is localised.

It is perhaps even more potent when it is in the form of tiny insoluble dust particles, minute quantities of which may cause lung cancer. Unlike the effects of potent chemical poisons, which take effect immediately, the cancers caused by plutonium are not likely to develop for many years.

Plutonium resembles radium in its toxicity and mode of action. The ICRP and the U.K. Medical Research Council recommend less than a millionth of a gram of plutonium-239 as the maximum amount that should be retained in the body of any person working with the substance.

Several witnesses suggested that this permissible level is too high, basing their argument on the so-called 'hot particle hypothesis'. This postulates that, if plutonium is localised in the form of a highly concentrated particle, particularly in the lung, it will be very much more likely to cause cancer than if it is uniformly distributed. It was argued that, since the hypothesis might be true, caution should be exercised in decisions concerning plutonium. No experimental evidence was presented in support of the hypothesis.

Many witnesses expressed concern at the potential long term polluting effects of low activity wastes containing small quantities of plutonium. These wastes, from reprocessing plants, are disposed of in the sea or by burial on land. Plutonium is also one of the long-lived components of high level waste. Reference has been made to the danger of its selective absorption by living organisms in certain circumstances, but the evidence does not permit us to draw a conclusion as to the extent of this danger.

Other witnesses claimed that the properties of plutonium are well understood, that the technology for its processing has been in use for more than twenty years and that there is still no published evidence which demonstrates a higher than normal incidence of lung or other cancers in people working with plutonium.

It was stated that some 5 kilograms of plutonium would be sufficient to poison every human being alive today, if it were distributed in the most lethal conceivable way. Such a result could not occur in an accident situation. Nevertheless, accidents could conceivably happen which would lead to dangerous dispersal and inhalation by large numbers of people. The possibility of plutonium being deliberately dispersed in the atmosphere is discussed in Chapter 14.

Hazards from uranium mining and milling

Mining

Uranium-bearing ores contain, in addition to uranium, smaller quantities of all the elements produced during its radioactive decay. These include radium and its decay product, radon, mentioned earlier. Breaking up the ore during mining facilitates release of the radon gas. The evidence shows that radon poses the principal radiation hazard in uranium mining. Its solid, alpha-emitting decay products, which can be retained in the lung, are the direct cause of the hazard.

Adequate monitoring is essential to ensure that miners are not exposed to radon concentrations above the permissible limits. Exposure is normally controlled by either natural or artificial ventilation. The evidence before the Commission indicates that natural ventilation is normally sufficient to maintain radon levels below the permissible limit in open-cut mines. Continual monitoring to ensure that particular atmospheric or other situations do not lead to high levels was recommended by witnesses and is required by the *Code of Practice*.

24647/76—7

Dust generated in the mining processes can also be a potential radiation hazard if it is inhaled and lodges in the lung. Control of dust is necessary to prevent exposure to hazardous levels of silica as well as radiation, and the limits prescribed for silica concentration generally reduce radiation exposure to well below prescribed levels. This may not be the case with very high grade ores, none of which have been found so far in the Ranger deposit.

The other possible radiation hazard encountered during mining is the dose received externally from radioactive decay in uranium ores. The evidence indicates that this is not a problem, except in the neighbourhood of very high grade ores. In such circumstances, the dose can be controlled by restricting the time that individual miners spend in high radiation areas.

Non-radiation hazards of the kinds faced in other mining operations are, of course, also present. These include risks associated with machinery, explosives, falling rock and so on. The Commission was not advised of any accidents that may occur in uranium mining which could create a significant radiation hazard.

Yellowcake production

Radon and dust can also present radiation hazards during milling, but they are likely to be less of a problem than in mining. Dust can normally be controlled satisfactorily, but radon levels should be monitored to check that the allowable limits are not exceeded. The radiation dose received by workers can be controlled, if required, by limiting the amount of time spent in any higher dose-rate areas.

Yellowcake, itself, is only slightly radioactive, as nearly the entire burden of uranium decay products is removed from the ore during milling. It presents no special problems, provided it is treated as a toxic substance. However, quantities stored in bulk may deliver radiation dose rates requiring limitations on access to storage areas.

Transport of yellowcake

Yellowcake is packed into metal drums, the dose rate at the surface of which is typically some thousandths of a rem per hour. The drums are normally transported in accordance with the requirements of the International Atomic Energy Agency's *Regulations for the Safe Transport of Radioactive Materials*. These regulations ensure adequate safety for the driver and for members of the public.

Disposal of tailings

The evidence suggests that the main hazards associated with yellowcake production arise from the material, known as tailings, left over after the uranium is extracted from the ore. This material contains all the radioactive decay products of the uranium, which were responsible for most of the radioactivity in the original ore. In the ore, these minerals were associated with a larger volume of non-radioactive rock. Milling converts all this material into a finely ground, more easily dispersed form.

One of the decay products, thorium-230, has a half-life of about 76 000 years. It decays into the radio-toxic nuclides radium-226, radon-222 and radon decay products, which will be continuously produced in dwindling amounts until all the thorium-230 has decayed away. With the uranium extracted, the concentrations of radium and radon will eventually fall to insignificant levels. However, this will not occur for more than 100 000 years, during which time the hazard will persist.

Several witnesses suggested that, over this period, dispersion of radon from a tailings storage could increase the incidence of lung cancer in generations of people living even at remote distances from the storage. At the same time, it was recognised that the widespread but variable distribution of naturally-occurring radon would have a similar effect and that differences in the incidence of lung cancer resulting from this natural variation appear to be undetectable against the background totality of lung cancer. No witnesses argued that an increase in radon release would have no effect on future human populations, but some argued that the risk from the increased level of radon from tailings was insignificant compared with the variations in risk arising from the differences in naturally occurring levels.

The radioactive decay products trapped in the tailings will emit gamma radiation, though at a diminishing rate, for as long as radium is present. Control of access to a tailings retention system reduces the hazard from this source during the life of a mine and milling operation. Long-term protection based on restricted access is obviously impracticable.

The extent to which the tailings from any milling operation become a danger to health will depend on the system adopted for their management. Return to the pit from which they were extracted is one possibility. In contrast to the 'tailings piles' remaining at many uranium mines in the U.S.A., the Ranger proposal involves storing the tailings behind an engineered dam. A risk with this type of storage is that tailings will escape to the environment if the dam embankment is breached.

During the operation of a mill, the tailings would normally be covered by a layer of water which restricts the release of radon. However, the water is not likely to remain after mining ceases, and continued restriction of radon release depends on the long-term efficacy of any layer of soil or other material which covers the tailings.

Radioactive material from tailings storage systems may be released to the environment by seepage and surface runoff. Another possible route is wind-borne dust that could be raised if tailings are allowed to dry out and remain uncovered. Because of the many variables inherent in both the design of tailings retention systems and in their local physical environments, the potential radiation hazards of each individual system should be assessed separately. An assessment of the Ranger proposal for the disposal of tailings will be included in the Second Report.

Hazards from heavy metals

Regions bearing commercial grade uranium ore also tend to contain greater or lesser amounts of toxic heavy metals such as lead, zinc, copper and cadmium, usually as sulphides. If released from mining or milling wastes in significant quantities, these metals would go into solution in due course and might do considerable damage to the local environment. The risks will be different at each mine, so each case calls for individual study before management practices are determined. Many witnesses stressed that monitoring is essential, both before and during mining operations.

Conversion of yellowcake to uranium hexafluoride

The principal hazards in the routine operation of this process arise from the toxicity of hydrogen fluoride and fluorine used in the production of uranium hexafluoride. Safe methods of handling these chemicals are established in the fluorochemical industry, which uses them in kilotonne quantities. Radiation hazards are very small. They are largely concentrated in the stream containing the small quantity of high activity contaminants extracted from the yellowcake in the purification process.

The product of conversion is very pure uranium hexafluoride, which is a highly corrosive gas as passed through enrichment plants, but a solid at room temperature. It can be inexpensively and safely packaged in steel cylinders.

It is at the point of discharge from the conversion operation as uranium hexafluoride that material in the fuel cycle passes into the NPT safeguards system set up by the IAEA (International Atomic Energy Agency). Under these safeguards, nuclear material in all subsequent operations of the cycle must be physically accounted for with the greatest precision feasible. The effectiveness of the safeguards system is discussed in Chapter 13.

Enrichment

The main potential hazard in the normal working of enrichment plants is accidental release of uranium hexafluoride. An important guard against this is the fact that most of the enrichment in diffusion plants, and all of it in centrifuge plants, is carried out at sub-atmospheric pressures. Feed and withdrawal stations, where uranium hexafluoride is sometimes at pressures substantially above atmospheric, are designed to contain any accidental releases. These stations also have to be physically separated from the enrichment areas for safeguards and inventory accounting purposes.

A potential cause of release of uranium hexafluoride in a centrifuge plant is an accident in which a rapidly-spinning centrifuge breaks up. There is a risk of this sort of accident causing a wave of destruction in a bank of centrifuges.

Because of the newness of the technology and the scarcity of published information, it is not possible to reach firm conclusions on the nature and seriousness of the hazards associated with centrifugal enrichment. But there is no evidence to suggest that the technology poses any particularly dangerous hazards in normal operation. For diffusion plants, the evidence suggests that only hazards of types common to large scale chemical industries will be encountered. However, as the level of enrichment increases, so too does the risk of accidentally bringing together enough uranium-235 to set off a chain reaction. Great care is needed to ensure that such an event never occurs.

The depleted uranium residue from enrichment plants is normally stockpiled for possible future use as a fertile component of reactor fuel. This material is mildly radioactive, and gradually produces the much more hazardous nuclides radium-226 and radon-222. Production of these nuclides is very slow, being limited by the long-lived intermediate decay products uranium-234 (half-life 250 000 years) and thorium-230 (half-life 76 000 years). At present, any radiation hazard from the stockpiles is controlled by limiting access to the area.

Nuclear fuel production

The fuel now used in most power reactors consists of ceramic pellets of uranium dioxide contained in metal cylinders. A type of fuel that may become important in the future is mixed oxide (MOX), in which fissile plutonium oxide is mixed with natural or depleted uranium oxide to give a fuel equivalent to enriched uranium. Other kinds of fuel include the metallic uranium fuel elements used in the Magnox reactors, particulate fuel developed for High Temperature Gas-Cooled Reactors, and carbide fuels being developed to enhance the performance of some types of reactor, particularly fast breeders.

The production of uranium dioxide fuel elements is now a well-established procedure which seems to be free of appreciable hazards. The pressing and sintering of uranium dioxide to form the required strong, highly stable pellets are carried out under conditions aimed at ensuring that uranium dioxide fines and dust do not escape into air which people will breathe.

Manufacture of mixed oxide fuel in the form of ceramic pellets is a much more difficult matter. Hazards arise from the toxicity of plutonium and from the fact that the 'critical mass' of plutonium dioxide in which chain fission reactions can start up is only a few kilograms. Great care must be taken to prevent the introduction of a moderator, such as water, which could cause a chain reaction to occur in a smaller quantity of plutonium (or enriched uranium).

Plutonium's toxicity makes it essential that any procedure involving the metal or its compounds be carried out by operators working outside totally enclosed cells containing the plutonium. Appropriate ventilation and adequate monitoring are required. The risk of a criticality accident arises from the fact that some plutonium, as nitrate or oxide, would normally be held in store in a fuel factory to provide feed for the production of fuel.

The evidence before the Commission indicates that the normal operational hazards of mixed oxide fuel production, although serious, are not intractable.

Reactors

Power reactor accidents

Partly because the fuel in a thermal reactor is composed mainly of non-fissile material, it is not possible for the core of such a reactor to explode like an atom bomb. However, many witnesses were concerned at the possibility of major power reactor accidents spreading dangerous amounts of radioactive material over wide areas. There was widespread acceptance among witnesses that an accident affecting members of the public is likely to occur sooner or later. Nevertheless, the fact that there is no evidence yet of any member of the public having been harmed as a result of an accident in a commercial power reactor, despite many 'abnormal occurrences', is a pointer to the overall safety of the power reactors installed to date.

Since it was established in January 1975, the U.S. Nuclear Regulatory Commission has published information on events at nuclear power stations 'which involved a temporary but significant reduction in the level of protection'. In the first six months of 1975, there were seven such events or classes of events, affecting 22 licensed power reactors; in the six months up to March 1976 there were two, affecting twenty licensed power reactors. (A class of event means an incident, such as cracks in a cooling pipe, which occurs in a number of reactors of the same design.) None of these occurrences had 'an actual impact on or consequence to the health and safety of the public'.

Abnormal occurrences at nuclear power stations, caused by equipment malfunctions and operator errors, have included partial cooling failures, such as occurred at the Dresden 2 BWR near Chicago in 1970, and a major fire at the Browns Ferry 1 and 2 BWRs in Alabama. Among non-commercial reactors, one of the most serious accidents that has occurred was a large heat and radiation release in 1961 at the SL-1, a small U.S. prototype military power plant, which killed three operators. Another was a severe fire in a British military plutonium production reactor at Windscale in 1957, during which considerable quantities of radioactive material, notably an estimated 20 000 curies of iodine-131, spread over the surrounding countryside.

A study of the risk of accidents in light water cooled power reactors, directed by Professor Norman Rasmussen of the Massachusetts Institute of Technology and funded by the U.S. Atomic Energy Commission, was referred to frequently in evidence. This *Reactor Safety Study* was published in 1975. Its findings indicate that, the more serious a reactor accident, the lower is its probability.

The most serious accident envisaged would be caused by the melting of a reactor core, releasing the highly radioactive fission products. The study estimates that there is one chance in 20 000 per reactor per year of this happening. If it did happen, according to the study, the molten core would probably melt through the bottom of the reactor and the floor of its concrete containment, eventually coming to rest in the ground some distance beneath the power station. There it would act as a long-term source of radioactive contamination, the effect of which would depend on where the core settled.

The reactor's containment probably would not be breached, according to the study, and most of the volatile fission products would settle on cool surfaces inside it. However, a small fraction, together with some of the gaseous fission products, could be expected to leak out into the atmosphere, mainly within a few hours of the accident. The study concludes that, given appropriate emergency measures, it is unlikely that anyone would be harmed, but the situation would have to be closely monitored and controlled.

However, the study acknowledges the possibility, which it concludes is remote, that the containment might fail, releasing virtually all the volatile and gaseous fission products into the atmosphere. It puts the probability of this happening at one chance in a thousand million per reactor per year. The study estimates that, if it did happen in a reactor of about 1000 MWe capacity situated on the worst possible site with maximum releases of fission products in the most adverse weather and wind conditions, 45 000 to 50 000 people in the surrounding population would suffer radiation illnesses soon after the accident. Most of these people would recover completely, but about 3300 would die. About 10 years after the accident an increased incidence of cancer would start to occur, eventually leading to about 45 000 fatal cancer cases spread over the ensuing 30 years or so. About 170 genetic defects per year would show up in the first generation, the number decreasing in subsequent generations.

A number of causes of a core melting can be postulated. One frequently referred to was a break in the pipes carrying the coolant, combined with the failure of emergency cooling. Evidence was given describing scale model tests in which light water reactor emergency core cooling systems (commonly abbreviated to ECCS) failed to operate. An incident in which the ECCS of a full-size reactor did not work when inadvertently set off was also mentioned. However, it was pointed out that the *Reactor Safety Study* found that the overall risk of a meltdown was increased by a factor of less than ten if the ECCS failed completely during a large loss of coolant accident. The reason for this low increase in risk is that the likelihood of a loss of coolant being accompanied by the other malfunctions necessary to cause a core meltdown was assessed as very remote.

The study used the 'fault tree analysis' method of risk assessment, in which hypothetical accident sequences are traced through all operations of the process or plant being assessed. Probabilities of failures are estimated at each step. The study was criticised by some witnesses who argued that it underestimated the risks of accidents. Critics also questioned the validity of the fault tree method. The arguments over the adequacy of the study are extensive and complex, and could not be fully examined in evidence before the Inquiry.

The study concluded that human error contributed to the overall risk, and that this factor could be assessed; such assessments were included in the study's risk estimates. Some witnesses raised the possibility that the level of operator competence might fall as the nuclear power industry grows, increasing the risks. Others argued that, as in the air transport industry, longer experience can be expected to lead to greater safety.

A common criticism of the study was that no account was taken in its risk assessments of the possibility of deliberate sabotage causing an accident. The study team excluded sabotage because of the impossibility of predicting precisely what part of a reactor might be sabotaged. But the study report said that all the possible events following an act of sabotage which could lead to a large release of radioactivity were examined, and concluded that sabotage by a small group could not cause more than 'largely superficial disruptions with only economic implications'. The risks of damage to reactors due to sabotage are discussed in Chapter 14.

Another type of risk not considered by the *Reactor Safety Study* is consequences of war. If a country with nuclear power stations were invaded, the stations might be abandoned in panic by their operators, with unforeseeable results. The strategic importance of power stations and the very large radioactive inventory of a reactor core led some witnesses to suggest that the risk of bomb attack during a war would be considerable. However, others argued that, because of the strength of the containment, dispersal of the core would need a near-direct hit by a large nuclear bomb; they said that, if a hostile power wished to maximise the radiological effect of a nuclear attack, it would be unlikely to target its weapons on nuclear power stations. The Commission concludes that a hostile power might, nevertheless, choose to attack nuclear power stations for other strategic reasons and that such an attack might result in a release of radioactivity from the reactor. From the evidence available to the Commission, it is not possible to determine whether such an attack would have greater radiological consequences than detonation of the nuclear weapons over other targets, such as large centres of population.

Current commercial power reactors are subject to government licensing and regulatory control at all phases of design, construction and operation. The Commission concludes from the evidence before it that current light water reactors are safe to a high degree of probability, and are likely to result in a lower total number of deaths per unit of energy than coal-fired power stations in this generation of people.

However, the evidence suggests that the long-term effects of ionising radiation from other stages of the nuclear fuel cycle may offset this comparative advantage, particularly if uranium mill tailings are not disposed of in a way that prevents the release of radon.

Outside the U.S.A., information about abnormal occurrences at, and the risks of accidents to, power reactors is less readily available. The Commission received insufficient evidence to enable it to reach a conclusion about types of reactor other than LWRs. However, it notes the arguments that CANDU reactors are less likely to suffer certain types of accident involving pressure failure of the coolant system. Pressure tube construction, as in the CANDU, means that a single failure in the pressurised pipework is less likely to lead to a major release of coolant.

Risk philosophy

Some witnesses stated that the risk, however small, of a catastrophic accident would be viewed by society differently from the cumulative risks of a number of much smaller accidents with the same total mortality, health and damage effects. It was also suggested that imposing the risks of the nuclear industry on society cannot be justified by claiming that society already accepts other, greater risks. Witnesses pointed out that society is working to reduce other society-imposed risks, such as those of air pollution and dam failures associated with other methods of power production. They argued that this shows society does not accept these risks, and certainly cannot be assumed to accept the new set of risks associated with nuclear power.

Witnesses criticised arguments for nuclear power based on comparisons of its risks with much greater risks to which people voluntarily expose themselves, such as those of driving a car. They argued that self-imposed risks could not be validly compared with those imposed by society. Also, the possibility was raised that many small increases in risk from different sources might accumulate to become a major increase in society-imposed risk.

Fast breeder reactors

The suggestion was made that liquid-metal cooled fast breeder reactors are inherently more dangerous than thermal reactors. The major concern arises from the theoretical possibility that an FBR core could form a critical configuration if it melted. Other factors suggested as possibly making them more dangerous are the use of sodium as coolant, the higher energy density, the greater neutron flux and the higher working temperature.

The evidence indicates that even a low order nuclear explosion in a fast breeder reactor is very unlikely, although it cannot be ruled out as a possibility. No commercial fast breeders have been built yet, but prototypes are delivering electricity to grids in several countries.

The potential economic advantage of fast breeder reactors derives from their ability to convert uranium-238 to plutonium-239 at a faster rate than they consume plutonium-239 as fuel. The additional fissile plutonium can then be used as fuel for other reactors. Fast breeders could convert a large proportion of the abundant uranium-238 to plutonium which then would be available as nuclear fuel.

A key to the whole process is the reprocessing of spent fuel rods, the extraction of plutonium, and its recycling as fuel. Thus plutonium recycling, which the evidence suggests may be economically optional for a program of thermal reactors, is an essential part of any fast breeder program based on uranium-238 as the fertile nuclide. Many witnesses were opposed to the development of fast breeder reactors because of the hazards involved in handling and transporting large quantities of plutonium. (See Chapter 8 for an assessment of the possible timetable for the introduction of fast breeder reactors and their likely impact on demand for uranium, Chapters 12 and 13 for discussion of the risk of diversion of plutonium by governments for nuclear weapons programs, and Chapter 14 for discussion of the risk of plutonium falling into the hands of terrorist groups.)

Earthquakes

The possibility was raised that damage to a nuclear power station caused by an earthquake would pose a serious hazard. Conceivable effects of an earthquake include breaching of the reactor containment, loss of coolant due to a breakage in the cooling circuit, damage to safety devices such as emergency cooling systems, and a breakdown in the mechanism that controls the reaction rate and normally shuts the reactor down.

Clearly it is essential that a system designed to shut down a reactor in an earthquake should remain intact. Several emergency systems have been devised for stopping the reaction if the mechanism that drives the control rods fails. One of these, used with light water reactors, is designed to automatically inject water containing boron compounds, which absorb neutrons, into the reactor core in an emergency. Another, installed in a Japanese gas cooled reactor, provides for boron-steel balls to drop into the core if normal shut-down systems fail.

Siting reactors away from areas prone to earthquakes is obviously desirable. The maximum probable seismic disturbance should be assessed for any site under consideration, and a plant for the site should be designed to survive such a disturbance with a large safety margin. It must be recognised, however, that the occurrence and characteristics of earthquakes are unpredictable. It follows that their effects are also unpredictable and that totally 'earthquake proof' designs are not practicable.

The U.S. Advisory Committee on Reactor Safeguards is at present re-examining its design policy for coping with earthquakes. The favoured practice has been to build flexible structures, so that they should bend with the shock. Now, particularly where seismic disturbances may be relatively severe, it is being suggested that design principles based on creating rigidity and strength should be adopted for foundations, concrete containments, floors and support structures.

The U.S. *Reactor Safety Study* considered, in its risk assessments, two specific reactors east of the Rocky Mountains where seismic activity is relatively low. It concluded that earthquakes did not make a significant contribution to the overall risk. The study has been criticised for not assessing the risk from earthquakes in regions of high seismic activity, such as the Pacific Coast of the U.S.A.

Routine operation of nuclear power stations

Reactors produce a variety of low activity wastes during their normal operation. The solids among them are generally buried, the liquids discharged into streams or the ocean, and the gases released into the atmosphere. The evidence suggests that present routine releases to the environment are well within permitted levels. However, some witnesses expressed concern that the low activity wastes from the greatly increased numbers of nuclear power stations officially predicted could at some stage add up to a considerable hazard.

Radioactive gases routinely released include argon-41; the activity of this gas released from one reactor in a day can amount to hundreds of curies. Because of the rapid decay of the gas (half-life 1.83 hours), discharge from a high chimney stack is regarded by control authorities as a satisfactory method of disposal.

The cooling circuit of a reactor accumulates radioactive material. Fuel rods may leak, releasing fission products, including gaseous krypton-85 and iodine-131. The fuel cladding becomes radioactive, and corrosion will cause some of this material to be released. Some of the atoms of water coolant, particularly heavy water, are converted by absorption of neutrons to radioactive tritium. Some of this contamination is released to the environment. The gases are diluted and vented to the atmosphere, sometimes after a period of storage to allow decay and consequent decrease in activity. Liquid wastes may also be released after similar treatment.

With careful monitoring and supervision, there seems little doubt that low level waste discharges can be kept well within presently permitted levels. If numbers of power reactors increase greatly, these levels may have to be reduced. The Commission was told that methods for removing most of the radioactivity in gaseous emissions from nuclear power stations have been developed.

Only about one-third of the heat energy produced in a power reactor core is converted into electricity. The remainder is waste heat and is rejected to the environment. The quantity of heat is very great; if it is rejected to a river or lake it may, depending on the location of the power station, raise the water temperature by several degrees Celsius, with very marked effects on the ecology of the water concerned. This phenomenon is known as thermal pollution. Fossil fuelled power stations also cause thermal pollution. However, for the same size stations, the effect of a nuclear station (other than a high temperature reactor or fast breeder) is significantly greater. The problem of thermal pollution is expected to increase the difficulty of finding suitable sites for power stations in years to come, particularly in countries which are not able to build them on the coast and discard the waste heat into the sea.

During normal operation of a nuclear power station, workers employed at the station are protected from any radiation hazard by the heavy shielding around the reactor core. Occasional accidents during maintenance or repair operations have led to workers receiving more than the permissible maximum dose of radiation. Workers have also been injured from time to time in non-radiological accidents. However, the overall safety record of the nuclear industry is a good one. British statistics show that the accident rate to nuclear power workers is less than the average accident rate for British industry as a whole.

Dismantling reactors

The write-off period for the financial cost of a power reactor, usually fixed at 20-30 years, does not necessarily reflect its useful life. This may be longer or shorter, depending largely on the relative costs of operating old and new equipment.

Complete dismantling of a nuclear power station will be difficult and hazardous, because of radioactivity induced in the reactor structure during its operating life. Bombardment by neutrons of the materials used to build a reactor produces a range of radioactive nuclides. Some of these emit highly penetrating gamma radiation, and have half-lives of several years.

The Commission was told that there is now some optimism about the possibility of totally disposing of power reactors. This follows experience gained in coping with mechanical problems inside reactor core structures, and in the successful dismantling of a small power reactor. It had been suggested that complete dismantling and disposal might cost as much as the original reactor, but recent estimates suggest that the cost may be much lower, perhaps as little as 5 percent of the original cost. However, until there is actual experience in disposing of a large power reactor, it will not be possible to estimate the cost precisely. The disposal cost must affect the economics of power generation; it should be a charge on the operator of the station concerned.

Remote handling behind temporary shielding will be required for dismantling operations. The scrap would presumably be disposed of in the same manner as other solid radioactive waste, probably by burial.

Storage of spent fuel

The spent fuel elements removed from reactors at refuelling are the most intensely radioactive material in the fuel cycle. The main hazard is the enormous amount of gamma radiation emitted by decay of radioactive fission products. The spent elements, handled by remote control from behind heavy shielding, are removed to deep tanks of water known as 'cooling ponds', and left there for some time. It is necessary to store them in the ponds in a way that prevents the considerable amount of fissile material present—uranium-235 and plutonium—from forming a critical configuration.

After the short-lived fission products have decayed to low levels of activity—which takes a few months—the fuel can be reprocessed. But at present, apart from the magnox fuel used by the first generation of nuclear power stations in the U.K. and France, very little fuel from power reactors is reprocessed in the Western world. Evidence before the Commission indicates that reprocessing of the oxide fuels used in the vast majority of reactors will not begin on a large scale in the U.S. or Europe before the mid-1980s, if then.

The number of spent fuel elements in storage is growing rapidly, and evidently will continue to do so for some time. Cooling ponds are satisfactory for short-term storage, but clearly they cannot be a permanent resting place for spent fuel. The ponds require continual surveillance and, despite the reduction in radioactivity during storage, the actinides in the spent elements will remain dangerously radioactive for hundreds of thousands of years.

Transport of spent fuel

Even after 'cooling' for several months, spent fuel can only be handled and transported safely when adequate shielding is provided to absorb radiation. Only approved containers may be used. To secure approval, the design must fulfil stringent specifications as to the effectiveness of the shielding, cooling arrangements and so on. Also, a prototype must have passed exacting mechanical and fire tests aimed at ensuring that the integrity of the package will be maintained in the face of any hazard encountered during transport.

Because of the need to fulfil these requirements, containers designed and approved for the transport of irradiated reactor fuel are large, weigh tens of tonnes, and are very ruggedly constructed. They can only be moved with heavy duty gear. Broaching of such containers, other than by remote handling behind heavy shielding, would release radiation doses so high that they would be quickly fatal. Spent reactor fuel, if unshielded, is so dangerous that it is effectively self-protecting against illicit seizure and diversion.

Reprocessing At a reprocessing plant, the spent fuel is dissolved in strong nitric acid and then separated into various components. Shielding is required to protect workers against the very high radiation levels. The radiation also tends to accelerate corrosion of plant components by the strong acid.

Reprocessing of magnox fuel, in which the uranium is in metallic form, is a well established technology. But the oxide fuels of modern reactors create greater problems in reprocessing. An oxide fuel element typically produces nearly ten times more energy per unit mass than a Magnox element—and therefore nearly a ten times greater quantity of fission products —before it is replaced by fresh fuel in the reactor. At present no major oxide fuel reprocessing plant is operating. Engineering problems, difficulties with licensing, and economic considerations have combined to bring this situation about.

The potentially useful products of reprocessing are uranium and plutonium. The uranium extracted from spent fuel from light water reactors is richer in uranium-235 than natural uranium and is desirable feedstock for enrichment plants. The uranium would go as uranium dioxide to a conversion plant before enrichment; such a recycling operation presents no new hazards. The hazards of fuel production using plutonium have been described.

Low and intermediate level wastes The gaseous fission products contained in the fuel elements, notably krypton-85, iodine-131 and iodine-129, are released from the fuel pellets during reprocessing. Tritium and volatile compounds of carbon-14 are also released. Much of the radioactive material is removed from the effluent gases, but krypton-85 (half-life 10.8 years) and tritium (half-life 12.3 years) are vented into the atmosphere from existing plants. Most of the output of these two gases will almost certainly have to be removed from stack emissions if radiation standards are to be maintained when large-scale oxide-fuel reprocessing begins. The Commission was assured that this could be done.

Low level liquid waste, eventually discharged to the environment, also arises in reprocessing. Generally, most of the small quantity of tritium is discharged with this waste. The evidence suggests that, in the future, storage or improved methods of disposal will be required.

Reprocessing plants also produce intermediate and low level solid wastes. Reprocessing the fuel used in one year by a 1000 megawatt light water reactor would produce some 20 to 60 cubic metres of such wastes. The main component of the intermediate level solids is fuel cladding material, radioactive to a degree depending on its composition and irradiation history. It is contaminated with small amounts of spent fuel. Low level solid wastes include contaminated laboratory apparatus, towels, rubber gloves and the like.

There was disagreement among witnesses about the size of the problem posed by the solid wastes. On the one hand, it was claimed that the actinides they contain would eventually build up to significant quantities and radioactivity levels. Other witnesses suggested that these wastes would cause little problem partly because their bulk could be reduced by compression and burning. Up to now, most waste of this kind has been buried on land or placed in canisters and dumped in the ocean. The Nuclear Energy Agency of the OECD is currently supervising the disposal of 7000 tonnes per year at a depth of 4500 metres in the Atlantic Ocean. Because of misgivings in the U.S.A. about burial and ocean dumping, disposal methods are being re-examined there. Disposal procedures are advanced in the Federal Republic of Germany, where low activity wastes are being disposed of in a massive rock salt formation at Asse.

High level waste

This intensely radioactive waste remains after uranium and plutonium have been extracted from spent fuel during reprocessing. Many witnesses discussed the disposal problems presented by the fact that some of its components take about one million years to decay to the general level of radioactivity in the environment. As an indication of the quantities involved, the Commission was told that reprocessing of the fuel used by one 1000 megawatt light water reactor in a year would yield about 30 cubic metres of liquid high level waste. Cumulative production up to the year 2000 may be over half a million cubic metres.

An indication of the radioactive content of high level waste can be obtained from a U.S. study. The authors, J. O. Blomeke, J. P. Nichols and W. C. McClain of the Oak Ridge National Laboratory, calculated the amount of waste that would be produced up to the year 2000 by a light water reactor power program that grew from a capacity of 5 GWe in 1970 to 1200 GWe in the year 2000. That is about 15 times the present world capacity, but rather less than the most recent projection of world capacity at the end of the century (see Table 5). If, in that year, all the wastes were uniformly dispersed through the huge volume of the oceans, the concentration of radioactive materials in sea water would be about 4 per cent of the maximum permissible concentration for drinking water. Because living organisms extract many radioactive materials from water and concentrate them, such a situation would not be acceptable. This example is purely illustrative and in no way represents what would happen if some high level waste did escape into the sea. On the one hand, it is inconceivable that all the waste could be released into the environment in this way. On the other hand, any that was would not be uniformly dispersed, but would be mainly confined to coastal waters and to sediments on the sea floor, so that the concentration would be correspondingly higher.

This waste is initially highly acidic. It contains all the fission products still active after the initial cooling, except the gases. Notable among these remaining fission products are strontium-90 and caesium-137, which take about 600 years to decay to one millionth of their originally very high levels of radioactivity. Strontium-90, if taken into the human body, deposits in bones and delivers much of its radiation dose to the vulnerable bone marrow. Actinides are also present, including small amounts of plutonium not separated in the earlier operations, and it is these nuclides which extend the danger period to hundreds of thousands of years.

The waste is at present stored as liquid in stainless steel tanks. However, this is only intended as interim storage; permanent arrangements have yet to be made. There was agreement among witnesses that the waste should be solidified for final disposal.

Various methods of solidification have been tested. They include allowing the radioactive liquid to boil itself dry leaving a cake-like solid, baking the waste into granules, and fusing it into dense glass. Some witnesses suggested that the structural integrity of solidified high level waste could not be guaranteed for more than a few decades. Others argued that tests in which glass loaded with actinides was exposed to very high alpha particle doses indicated that wastes fused into glass would remain stable for several thousand years.

It is anticipated that the high level waste from one year's operation of a 1000 MWe light water reactor, when solidified, would occupy a volume of about 2 cubic metres. One proposal is that this solidified waste be placed in cylindrical steel canisters prior to ultimate disposal; 2 cubic metres would fill ten canisters. These would be extremely radioactive, even 100 years after the spent fuel rods, from which the waste was separated, were removed from the reactor. Assuming that by this time the steel canister would have corroded away, Bernard L. Cohen of the University of Pittsburgh has calculated that a person who came within 10 metres of the waste from just one canister at this time, and was separated from it only by the intervening air, would receive a radiation dose of 100 rem per hour. He or she would receive the ICRP allowable annual radiation dose equivalent for whole body exposure for radiation workers in 3 minutes, and the allowable annual dose for a member of the public in 18 seconds.

For ultimate disposal, witnesses agreed that any reasonable possibility of the material being reintroduced into the environment within several hundred thousand years had to be precluded. None disputed that this rules out any type of permanent storage requiring supervision. Disposal of the waste as a solid in stable geological formations was the proposed solution seen as offering the best chance of permanent isolation. The main disagreement was over whether this really was a safe and practical answer.

The U.S. ERDA is at present conducting a search for suitable underground disposal sites. Some rock salt structures are regarded as promising, partly because they are expected to remain geologically stable and dry for a very long time and also because of the ability of salt to accommodate ground movements without permanent cracking. Under ERDA's program, underground disposal of solidified waste is planned to begin at about the end of the century, with storage under controlled conditions in the meantime. The Federal Republic of Germany is investigating the possibility of disposing of solidified high activity waste in the Asse salt formation.

One problem with underground storage is the impossibility of predicting with absolute certainty that any geological formation will remain stable for hundreds of thousands of years. Apart from natural processes, human actions, such as drilling or the setting off of underground nuclear explosions, could cause disturbances. This could happen at some future time when the presence of the wastes has been forgotten. There was disagreement among witnesses about the significance of these risks.

Another possible problem is the fact that any containment around the solidified waste is likely to corrode. It was argued that radioactive material could then find its way into underground water. The counter-argument was that corrosion of the containment would not cause significant release of radioactivity because of the low probability of water coming into contact with the solidified waste and the low rate of leaching if it did.

Apart from salt structures, geological formations regarded as possible permanent disposal sites include hard rock structures free of faults and fissures, and clay formations. One witness suggested that areas of very old, crystalline rock in central Australia were perhaps the most suitable sites in the world for disposing of high level waste. They have a history of well over a million years of geological stability, and have experienced a very dry climate for the past 10 million years. It was argued that, because of the suitability of these areas for waste disposal, Australia should consider reprocessing and disposing of nuclear wastes from other countries. Other witnesses questioned the suitability of the proposed disposal areas and rejected the idea that Australia might provide disposal facilities for wastes from other countries. An international study program on high level waste disposal in geological formations begins this year under the auspices of OECD.

A possible method of reducing from hundreds of thousands of years to centuries the length of time that high level wastes will remain dangerously radioactive was described in evidence. It involves separating the actinides from the fission products and then irradiating them in a reactor so that they undergo fission. The long-lived actinides would be replaced by shorter-lived fission products. This would not appreciably reduce the initial hazard, but after about five centuries the radioactivity of the waste would fall to something like one per cent of the activity that would have been present if the actinides had not undergone fission.

Many witnesses doubted the feasibility of this method. The fact that an actinide sample would have to pass through the irradiating reactor many times before all of it underwent fission, and the fact that the fission products would have to be separated out after each irradiation, were seen as major drawbacks. Doubt was also expressed about the likelihood of being able to achieve complete separation of the actinides and fission products.

Another suggestion put forward for the disposal of high level wastes is burial in canisters in the sea bed. The evidence suggests that this proposal needs much more research and at present appears less promising than burial on land. Firing the wastes by rocket into the sun and burying them in canisters in the Antarctic ice cap have also been proposed, and generally rejected as undesirable or not feasible.

Plate 5. A pressurised water reactor power station at San Onofre, southern California. The capacity of this station is 430 megawatts electrical. The reactor is housed in the steel containment sphere near the centre of the picture. More than half of the power reactors in the U.S. are PWRs; nearly all of the remainder are boiling water reactors. (Photo by courtesy of the U.S. ERDA.)

International agreements on radioactive waste disposal

A number of international agreements dealing with the disposal of radioactive materials beyond national boundaries exist. Some apply globally and others to specific regions.

Probably the most significant of the global agreements is the Convention on the Prevention of Marine Pollution by Dumping of Wastes and Other Matter, which was drawn up at an international conference in London in November 1972. This convention came into force in 1975, and at the end of that year twenty-two countries had become party to it, including the Soviet Union, the United States, the United Kingdom, Canada and New Zealand, but not Australia. Parties undertake, among other things, to prohibit the dumping at sea of high-level radioactive wastes and other high-level radioactive matter. The task of defining the material that the prohibition applies to is given to the 'competent international body' in the field, at present the International Atomic Energy Agency.

Under the convention, dumping of other radioactive material requires a special permit. All countries party to the convention are required to designate appropriate permit-issuing authorities, keep records of dumping activities, and participate in monitoring the conditions of the seas. The authorities issuing permits for dumping must take full account of the recommendations of the 'competent international body'.

In January 1975, the IAEA issued what were termed 'Provisional Definitions and Recommendations Concerning Radioactive Wastes under the Convention'. These include a definition of high-level radioactive wastes or other high-level radioactive matter unsuitable for dumping at sea; recommendations for the issuing of special permits for radioactive materials that may be dumped; and recommendations for the operational control of dumping of waste (selection of sites, packaging requirements, keeping of records, and so on).

Parties to the convention undertake to take measures to prevent and punish contravention of its provisions, and to co-operate in the development of procedures for reporting ships and aircraft found contravening them.

A more general convention which came into force in 1962, the Convention on the High Seas, contains an article requiring parties to it to 'take measures' to prevent pollution of the seas from the dumping of radioactive waste. Countries are required to take into account any standards and regulations formulated by the 'competent international organisations' in preventing pollution of the seas, or the air space above, resulting from any activities involving 'radioactive materials or other harmful agents'.

At the end of 1975 a total of 51 countries had ratified or acceded to this convention. However important its provisions may be in certain matters, its article relating to the dumping of radioactive materials has not had any impact on the development of the law in this area.

The whole question of the disposal of waste at sea is currently being considered at the Third U.N. Conference on the Law of the Sea. An entire chapter of the Draft Revised Negotiating Text under consideration at the conference is devoted to the 'Protection and Preservation of the Marine Environment'. The object of a convention based on the draft would be to require states to take all necessary measures to prevent, reduce and control pollution of the marine environment from any source.

Turning to regional conventions, the Oslo Convention on the Control of Marine Pollution by Dumping from Ships and Aircraft has been in force since 1974. It deals with dumping at sea in a region taking in the European waters of the Atlantic and Arctic Oceans. Although it does not refer specifically to radioactive material, its prohibitions on dumping can reasonably be interpreted as including categories of radioactive waste. A convention applying to the same region, the Convention for the Prevention of Marine Pollution from Land-Based Sources, does refer to radioactive material. Parties to this agreement undertake to forestall and eliminate pollution by radioactive substances finding their way into the sea from land-based sources.

Other regional agreements have been negotiated covering dumping or discharge by ships, and pollution from land-based sources, in respect of the Baltic and Mediterranean Seas.

The treaty establishing the European Atomic Energy Community (Euratom) contains provisions relating to radioactive waste disposal. Member countries are required to supply the Euratom Commission with information on any plans for disposal of such waste. The information must be sufficient to show whether implementation of a disposal plan is liable to result in radioactive contamination of the water, soil, or airspace of another Euratom member.

Any member country which considers that another member has failed to fulfil an obligation under the treaty may bring the matter before the Court of Justice established by the treaty. The members of Euratom are Belgium, Denmark, the Federal Republic of Germany, France, Ireland, Italy, Luxembourg, the Netherlands and the United Kingdom.

The disposal of radioactive waste in Antarctica is prohibited by the Antarctic Treaty, in force since 1961. This agreement was entered into by Argentina, Australia, Belgium, Chile, France, Japan, New Zealand, Norway, South Africa, the U.K., the U.S.A., and the U.S.S.R. It was an attempt to keep cold war rivalries out of the region, partly by putting into abeyance a number of competing claims to sovereignty over the region and partly by declaring that 'Antarctica shall be used for peaceful purposes only'. Both the disposal of radioactive material and nuclear explosions are prohibited in the region by the treaty.

The terms of recent conventions reflect a trend towards acceptance of the idea that states should be responsible for any damage arising from the disposal of wastes. The Convention on the Prevention of Marine Pollution by Dumping of Wastes and Other Matter, for example, provides that the parties to it will develop procedures for assessing liability and settling disputes regarding dumping. This, the convention specifies, will be done 'in accordance with the principles of international law regarding State responsibility for damage to the environment of other States, or to any other area of the environment, caused by dumping of wastes and other matter of all kinds'.

A number of conventions relating to civil liability for nuclear damage have been drawn up, and contain provisions dealing with responsibility for, among other things, damage from radioactive waste. Two are in force—the Paris Convention on Third Party Liability in the Field of Nuclear Energy, and the Brussels Convention Supplementary Thereto. They make the operator of a nuclear installation liable for radioactive damage caused by it or for which it is deemed to be responsible. The conventions set limits to an operator's financial liability.

24647/76—8

Plate 6. A fuel assembly is lowered into the core of the San Onofre pressurised water reactor shown in Plate 5. The assembly is a collection of metal tubes, about 4 metres long, containing nuclear fuel. (Photo by courtesy of the U.S. ERDA.)

Pressure towards the creation of a worldwide approach to the problems of pollution led the U.N. to call the Conference on the Human Environment, which met at Stockholm in 1972. This conference produced nothing concrete in the form of legally binding rules. Instead it formulated a number of very general principles which were backed up by a series of recommendations on possible action that would, if followed, implement those principles.

One of the principles expressed the conviction that the 'discharge of toxic substances or of other substances and the release of heat, in such quantities or concentrations as to exceed the capacity of the environment to render them harmless, must be halted in order to ensure that serious or irreversible damage is not inflicted upon ecosystems'.

One of the recommendations adopted was that governments should explore with the IAEA and the World Health Organization the feasibility of developing a registry of releases to the biosphere of significant quantities of radioactive materials. The resolution goes on to say that governments should support and expand, under the IAEA and appropriate international organisations, international co-operation on radioactive waste problems. The resolution specifically mentions problems of mining and tailings, and co-ordination of plans for the siting of fuel reprocessing plants in relation to the siting of ultimate storage areas.

Although there is an increasing number of international agreements covering radioactive waste disposal, they do not create any coherent set of international rules or standards. States are only bound by those treaties to which they are party. Hence, even an agreement as important as the Dumping Convention, which has as parties three major nuclear countries, is reduced in significance by the large number of states that have not ratified or acceded to it. Furthermore, to a greater or lesser extent, all these arrangements suffer from the fact that states are reluctant to allow other states to exercise jurisdiction over their ships or nationals. Hence, for the most part, enforcement of the legal proscriptions is a matter for each state to carry out for itself.

Estimates of general radiation hazard from nuclear power programs

Despite the inevitably large uncertainties involved, a number of estimates have been made of the contribution of nuclear power production to general radiation exposure. In 1972, for example, the U.S. Academy of Sciences and National Research Council published a report, by the Advisory Committee on the Biological Effects of Ionising Radiation under the title *The effects on populations of exposure to low levels of ionising radiation*, in which it is estimated that the average dose received by people in the U.S. from normally operating power reactors was 0.002 millirem in 1970, and would rise to 0.17 millirem in the year 2000. This takes no account of radioactive releases from other stages of the fuel cycle or of unplanned releases at any stage. These figures compare with an average dose from natural background radiation of 100 millirem per year.

A report by Sir Edward Pochin, entitled *Estimated population exposure from nuclear power production and other radiation sources*, was published in 1976 by the OECD Nuclear Energy Agency. The report looks at the total contribution made by all stages of the nuclear fuel cycle to the general radiation level. It concludes that nuclear power production at the rate of 1 kilowatt per head of population would give an average radiation dose to each person of 6 millirem per year—about six per cent of the average natural background. Australia's average per capita electrical power output is about half a kilowatt, none of it generated by nuclear power, so the production rate assumed in the study would apply in this country only if the average person doubled his consumption and all the electricity come from nuclear plants.

The OECD estimate includes a contribution from power station accidents, based on the risk assessments contained in the U.S. *Reactor Safety Study*. The author of the OECD report calculates that nuclear power production at the assumed rate of one kilowatt per head of population could cause about one fatal and one curable cancer case per million people each year. After some generations, 1 to 1.5 genetic defects per million people could be expected each year. Some witnesses, pointing to the genetic effects, questioned the right of people alive today to satisfy their demand for power by using technology harmful to members of future generations, however small the numbers affected.

Attitudes in countries with nuclear power industries

We have listed the countries with major nuclear power industries. We are not in a position to give a full account of the attitudes of the people in those countries to the presence of nuclear reactors. There is, we know, sizable opposition in the U.S.A. and in Japan, but we are not able to assess the exact extent of it and the degree to which any particular ground of opposition is supported. There is also some opposition in the United Kingdom and in the Federal Republic of Germany. The position of the governments in all the countries mentioned is clear; they support the maintenance of existing nuclear power production and plan expansion, although possibly with some reservations.

We think it is true of the position in those countries that nuclear power programs have developed as a result of commercial and governmental decisions, with little if any public scrutiny either of what was planned or of what had already been done. In Great Britain, a Royal Commission in 1974 undertook an inquiry into pollution aspects of the nuclear fuel cycle and is about to report. This may have an effect on public attitudes there.

The evidence before us does show that nuclear reactors appear to the observer, and to persons living nearby, to be relatively free of harmful environmental effects. They can be regarded as more pleasant to look at, and to live near, than fossil fuel stations. On present information, the operation of fossil fuel stations may fairly be regarded as constituting a greater risk to health. This assessment takes no account of risks of serious accidents in, or to, reactors, which are discussed elsewhere in this chapter. When it comes to other operations in the fuel cycle, the position is not so clear.

We do not have much information on the attitude of workers in nuclear power plants and other facilities in the nuclear fuel cycle, but in the U.S.A., the U.K. and the Federal Republic of Germany it seems that stringent safety precautions are taken, and workers seem to accept the conditions.

We do not attempt to draw any conclusions from the recent Californian referendum respecting nuclear reactors. We know enough about it to suggest that anyone who wishes to draw any conclusions should examine the situation carefully, and look particularly at the issue as formulated and the nature of the arguments presented on both sides.

Conclusions

Radiations are emitted at all stages of the nuclear fuel cycle. It is believed that they have harmful biological consequences which, in general, increase in direct proportion to rising dose. Apart from rapidly fatal effects (within days) of high doses, the major consequences of radiation are cancer which may occur some years after the individual has been exposed and gene mutations which may appear in subsequent generations. Internationally defined dose limits are set down in legislation and codes of practice in some countries to minimise harmful long-term effects for workers in the nuclear industry.

With the exception of some uranium mines, the safety record of the nuclear industry has been good. The accident rate to workers in the industry is no higher, and may be lower, than the average accident rate for industry as a whole.

A higher than normal incidence of respiratory cancer has been found among underground uranium miners in some parts of the world. This is attributed to the inhalation of air containing radon gas and its short-lived decay products. The risks to workers in open-cut mines, of the type envisaged in the first stage of the Ranger proposal, are lower. Under conditions set by the Australian *Code of Practice*, exposure of workers is to be monitored to ensure that total working-life doses are below what has been observed to result in an increased frequency of cancer.

Hazards to members of the public and the general environment arise at all stages of the nuclear fuel cycle. However, few problems seem likely to arise from conversion, enrichment or fuel fabrication plants. The main concerns arise from uranium mill tailings, reactors, the wastes from spent fuel reprocessing, and any situation involving plutonium.

Tailings contain radioactive materials which will remain harmful for over 100 000 years. Any health effects that arise if these escape to the environment will be suffered largely by generations remote from those who receive the benefit of the energy obtained from the extracted uranium. It is essential that the chance of such health effects occurring be minimised by appropriate engineering methods of containing the tailings.

The probability of a serious accident occurring in nuclear power stations is much debated. Thermal nuclear power reactors operating commercially have a good safety record. So far, there has been no dangerous release of radiation nor evidence of serious harm to the public resulting from radiological exposure. As a result of predictive studies on light water reactors in the U.S., the risks of dangerous events such as a core melt-down are estimated to be very small. The more serious the accident, the smaller is its predicted probability of occurrence. Imponderables not accounted for in risk estimations include deliberate sabotage and consequences of war. Since the occurrence and characteristics of earthquakes are unpredictable, their effects are also essentially unpredictable. The conclusion with regard to thermal reactors is that there is a very small but finite probability of a serious accident with release of highly dangerous radioactive material. The numerous incidents involving faulty function that have occurred in nuclear power stations, together with serious accidents known to have occurred in military installations, give no grounds for complacency.

There are theoretical reasons why fast breeder reactors may be potentially more dangerous than thermal reactors, the most important being that a core meltdown could conceivably lead to the development of an inefficient nuclear explosion. As yet there is insufficient practical experience with fast breeders to enable firm conclusions to be drawn. A more serious overall hazard associated with fast breeder reactors arises from the fact that their large-scale introduction requires that spent fuel be reprocessed for the extraction of plutonium and that the plutonium be recycled as fuel. Plutonium is highly radio-toxic, having the potential to induce cancer if it reaches certain organs even in minute quantities. It can also be used in atomic weapons. Its production and use on a large scale will considerably increase the dangers associated with nuclear power.

During normal operation, nuclear power stations produce radioactive gaseous and liquid wastes of low activity, and waste heat. At present these types of pollutant are released into the environment at levels which can be controlled within permitted discharge limits. Increase in numbers of nuclear power stations will necessitate more rigorous control of routine releases in order to avert cumulative harmful environmental consequences. At the end of their lives, reactors will have to be dismantled or otherwise disposed of. Parts of the reactor structure will be highly radioactive and their disposal could be very difficult. There is at present no experience of dismantling a full size power reactor.

After removal from the reactor, spent fuel, which is extremely radioactive, is stored under water in tanks. Subsequent transport of the still highly active spent fuel to a reprocessing plant requires specially constructed and shielded containers.

At present most spent fuel is being held in storage tanks, as no commercial plants for reprocessing oxide fuel are operating. Reprocessing of spent nuclear fuel yields radioactive wastes of varying activity. Low-level wastes are usually released into the environment. Intermediate wastes in solid form are either stored, buried on land or dumped under international supervision in the deep ocean. It is generally agreed that present methods used for burial or ocean disposal will have to be improved if these procedures are to be a satisfactory long term solution. High level wastes are at present stored mainly in liquid form, and some constituents will remain dangerously radioactive for several hundreds of thousands of years. There is at present no generally accepted means by which high level waste can be permanently isolated from the environment and remain safe for very long periods. Processes for the conversion of high level waste to a relatively inert solid are being developed. Permanent disposal of high-level solid wastes in stable geological formations is regarded as the most likely solution, but has yet to be demonstrated as feasible. It is not certain that such methods and disposal sites will entirely prevent radioactive releases following disturbances caused by natural processes or human activity.

Ocean disposal of high-level radioactive wastes has been extensively restricted by international and regional conventions which are binding on many countries with nuclear power industries. Disposal in Antarctica is prohibited by Treaty.

11 ENVIRONMENTAL HAZARDS OF NON-NUCLEAR ENERGY SOURCES

A complete assessment of the environmental hazards of nuclear power would require that these hazards be compared with those of alternatives to nuclear energy. A detailed comparison is impracticable because of the different nature of the hazards and because there are many alternatives to the use of uranium for electricity generation (see Chapter 6). The most common alternative is coal-fired generation of electricity. Oil-fired and natural-gas-fired power stations are unlikely in the future to be chosen as base load stations in place of nuclear stations, except perhaps in a few special cases, because of the relatively high fuel costs involved.

As noted in Chapter 5, projections for developed countries generally envisage that the proportion of energy consumed as electricity will increase. However, as an alternative, direct combustion of coal, oil or natural gas could be increased; for example, oil-fired rather than electric central heating could be encouraged. Direct combustion of fossil fuels could in this way be made an alternative to nuclear power. Such an alternative avoids one of the major disadvantages of electricity generation by the use of boilers and turbines, namely the wasting of about two-thirds of the heat produced in combustion. It substitutes other disadvantages, including air pollution at the point of use and transport and handling difficulties, particularly for coal.

It was suggested to the Commission that direct and indirect uses of solar energy could be substituted for nuclear energy in the total supply picture. This does not necessarily mean that solar energy would be used to generate electricity; for example, space heating by direct solar radiation could replace electric heating. At the same time, the evidence before the Commission suggests the desirability of a vigorous worldwide program of energy conservation, particularly in the developed countries. This recognises that at least part of the alternative to nuclear energy could be a reduction in the rate of use of energy, which must result in a reduction in environmental effects.

Major deleterious effects of non-nuclear energy technologies can be listed as follows:

Coal mining

- Accidents to miners (especially in underground mines).
- Pneumoconiosis and other diseases of miners (especially in underground mines).
- Drainage of water polluted by acid and heavy metals.
- Destruction of land surface by mine workings (especially from open cut mines). Restoration is normally carried out in some countries including the U.K. and the Federal Republic of Germany.
- Destruction of land and houses by spoil heaps.
- Ground subsidence (from underground mines).

- Disruption of underground water supply and depletion of scarce water resources (applicable to dry areas such as the western U.S.A.).
- Intrusion of industrial development in areas of high amenity and wilderness value (applicable to potential new mines in areas such as the western U.S.A.).
- Depletion of a non-renewable resource.

Coal transport and processing

- Large, intrusive rail transport systems, with attendant accidents.
- Air pollution by wind blown particles and by smoke and chemicals from coal processing plants.
- Water pollution from coal washeries and process plants.

Oil and gas production

- Pollution of sea and shore by spilled oil (from off-shore oil wells).
- Large scale industrial development for the construction of drilling rigs and other apparatus, often at coastal sites of high amenity value.
- Depletion of non-renewable resources.

Oil transport and refining

- Pollution of the sea and shore by planned or accidental spillage of oil from tankers.
- Fire and explosion on tankers, especially if in harbour at the time (particularly hazardous in the case of liquid natural gas tankers).
- Heavy road traffic caused by transport of refined oil products.
- Intrusion of tank farms, pipelines and refineries, often in coastal areas of high amenity value.
- Water pollution from oil refineries.
- Air pollution by sulphur dioxide and odoriferous gases from oil refining.

Combustion of coal, oil and natural gas

- Air pollution by sulphur dioxide, carbon monoxide and nitrogen oxides.
- Air pollution by small particles (applicable to combustion of coal); the combination of small particles and sulphur dioxide pollution can be a major health hazard, giving rise to problems that range from breathing difficulties to death.
- In the case of coal, creation of large volumes of residual material.
- Possible long-term climatic effects from carbon dioxide.

Effects of electricity generation in fossil-fuel-fired stations

- All the effects associated with the production, handling, transport and combustion of fossil fuels.
- Thermal pollution of water and air by waste heat discharge, including effects on local climate (see Chapter 5).
- Large, often unsightly buildings and transmission line pylons.
- Water pollution by heavy metals and other chemicals as a result of cooling water treatment and equipment cleaning.

Renewable energy sources

- Destruction of or deleterious effects upon areas of high amenity and wilderness value by hydro-electric, tidal and geothermal schemes.
- The risk that a hydro-electric dam might burst.
- Visual intrusion of large scale structures for wind power generation.

- Use of large areas of ground by large solar energy installations and consequent effects on local climate.
- Production of large quantities of the toxic metals lead, nickel and cadmium for storage batteries.
- Use of large areas of agricultural or forest land in the production of crops for fuel.

It should be emphasised that the consequences of most forms of environmental hazard associated with energy production and use can be and are being mitigated, whether the energy source is nuclear or otherwise. Examples from different countries are instructive. For example, the loss of fertile agricultural land as a result of strip mining of coal is a major problem in the Midwest coalfields of the U.S.A. However, in the U.K. and the Federal Republic of Germany it has been standard practice for many years to fully restore farmland after mining has ended so that this form of environmental destruction does not occur.

Among a sample of U.S. coal mining companies, the injury rate to underground miners was more than 20 times higher in the worst company than in the best. Underground fatality rates are two to four times higher in the U.S.A. than in the U.K. These figures suggest that there may be scope for reducing this particular environmental hazard simply by using currently available equipment and techniques.

Of the hazards listed, only some of those associated with mining and those directly associated with electricity generation are also hazards of nuclear power. One of the latter, thermal pollution of water, is generally considerably greater from nuclear power stations than from fossil-fuel-fired stations, because of the lower thermal efficiency of most nuclear stations (see Chapter 10). For the same energy production, the quantity of wastes resulting from uranium mining is considerably less than that from the mining of coal. However, uranium mines produce additional pollutants in the form of radioactive materials, such as radon (see Chapter 10).

It can be seen that the environmental hazards of fossil fuel use are severe, particularly in the case of coal, but that there is very considerable scope for reducing them. However, since the hazards of non-nuclear energy are almost entirely different in kind from the hazards of the nuclear fuel cycle, there is no simple way of comparing them. One significant difference is the fact that, while most of the potential adverse health effects of fossil fuel use accrue to the generation of people using the energy, any radiation effects from radioactive wastes and tailings are likely to be spread over many generations.

Conservation of energy confers very significant environmental benefits through the avoidance of hazards associated with energy production and use. The evidence indicates that a carefully planned conservation program need have no significant adverse effects on life-style or well-being.

Plate 7. This picture shows an irradiated fuel rod from an advanced gas-cooled reactor in the U.K. being examined from behind a thick protective shield. The fuel is handled by remote control. (Photo by courtesy of the U.K. Atomic Energy Authority.)

Plate 8. The U.K.'s 250 megawatt electrical prototype fast reactor (PFR) is in the foreground and the smaller Dounreay fast reactor behind it. The PFR started generating electricity in 1975. The reactors are situated on the far north coast of Scotland. (Photo by courtesy of the U.K. Atomic Energy Authority.)

12 SAFEGUARDS AGAINST DIVERSION TO WEAPONS-MAKING

The Commission was presented with several strong representations that the global dangers of any extension of nuclear power programs far outweigh any economic benefit that might accrue to Australia from the sale of its uranium. It was contended that international safeguards were inadequate to achieve their intended objective of preventing the diversion by governments to weapons-making purposes of nuclear materials, services, equipment, facilities and information used in a nuclear power program. In this Chapter, we identify potential weapons material and describe the safeguards systems now operating. The adequacy of the safeguards for preventing countries without nuclear weapons from acquiring them is discussed in Chapter 13.

Possible weapons materials

A nuclear explosion is the result of a rapid, uncontrolled chain reaction in a piece of fissile material, which may be plutonium-239, uranium-235, uranium-233, or mixtures of these nuclides. The piece needs to be larger than a certain mass, termed the critical mass. The mass is different for plutonium-239, uranium-235, and uranium-233; it is generally about three times greater for uranium-235 than for plutonium-239. It varies also with such factors as the purity of the material, its geometrical shape, density and temperature, and the material surrounding it.

Some fission explosive devices work by bringing together, at great speed, several pieces of fissile material of less than critical mass to form a single piece of greater than critical mass. Others depend on suddenly increasing the density of a subcritical mass by carefully designed implosion, using conventional high explosive, to form a super-critical configuration. A source of neutrons to initiate the chain reaction is also needed. Although stray neutrons from spontaneous fission or from cosmic radiation can initiate the chain reaction, a much greater and more predictable explosive effect (yield) can be obtained using a strong pulse of neutrons from a neutron source.

To obtain the maximum yield, the chain reaction must not be allowed to occur until the mass is fully assembled in its most super-critical configuration. Premature initiation of the reaction can cause the fissile material to blow apart before it is fully assembled, resulting in a low yield explosion. It requires some skill in bomb design and construction to ensure that this does not occur.

Successful bomb design requires a detailed knowledge of the properties of the particular fissile material used and of the behaviour of the conventional (chemical) explosive mechanism needed to bring the super-critical mass together. This knowledge used to be a closely-guarded secret, but today key information is openly available in technical publications.

The same is not so true of the thermonuclear or hydrogen bomb, most of the key aspects of which are still a military secret. The hydrogen bomb contains a layer of material, lithium deuteride, for example, which will undergo thermonuclear fusion. This is packed round a nuclear fission bomb. Detonation of the fission bomb generates temperatures and pressures so high that the nuclear fusion reaction is set off, greatly increasing the yield of the bomb as a whole.

To build a nuclear bomb, a supply of appropriate fissile material is essential. All the main nuclear weapon states obtained their supplies by means of special projects in military nuclear technology. The first nuclear reactors were built solely to produce weapons-grade plutonium, and the first enrichment plants were constructed to produce very highly enriched (about 90 per cent) uranium for bombs. The nuclear power industry developed from these projects; indeed, even today commercial enrichment capacity consists largely of plants originally built to produce highly enriched uranium for nuclear weapons. It seems likely that there will be a number of new enrichment plants within a few years, and this fact increases concern about nuclear proliferation.

Although at present uranium highly enriched in uranium-235 is only produced by the complex and expensive gas diffusion process, new technology (see Chapter 3) may well bring enrichment within the reach of many more countries. Also, uranium-233 can be produced from thorium-232 in high temperature gas-cooled reactors and breeders using the uranium-thorium system. The fuel for high temperature reactors may be very highly enriched uranium, which itself may be suitable for the production of bombs without further modification, depending on the actual degree of enrichment.

At present, plutonium-239 appears to be the isotope giving most concern about proliferation. It is produced, with other plutonium isotopes, from uranium-238 in all reactors using low-enriched or natural uranium fuel. Prototype fast breeder reactors use plutonium as fuel and, as explained in Chapter 3, it is intended that commercial FBRs will produce more plutonium than they consume. Plutonium is separated from spent fuel in reprocessing. Construction of small scale, possibly rather inefficient, reprocessing plants seems to be within the capacity of many countries; such plants now exist in the recognised nuclear weapons countries as well as in the Federal Republic of Germany, Japan, India, Italy, Spain, Yugoslavia, Argentina and Taiwan (see Chapter 4).

The rate of plutonium production varies with the type of reactor; for example, a CANDU heavy water reactor using natural uranium fuel produces about twice as much in proportion to the uranium-235 consumed as a light water reactor fuelled with slightly enriched uranium. Some of the plutonium-239 produced in a reactor core is converted by neutron capture to plutonium-240, the proportion of which increases the longer the fuel remains in the reactor. Atoms of this isotope do not readily undergo fission, when struck by a neutron, but have a fairly high spontaneous fission rate. Plutonium-241 and -242 are also produced in small amounts, likewise increasing with irradiation time.

The presence of these other isotopes mixed with plutonium-239 reduces the plutonium's quality as bomb material, both by dilution and because neutrons emitted by spontaneous fission of plutonium-240 heighten the tendency for the chain reaction to be set off before the super-critical mass is fully assembled. This makes reactor grade plutonium a less satisfactory material for bombs so far as a nation wishing to develop a nuclear arsenal is concerned, but it may be quite adequate for terrorist purposes (see Chapter

14). Such a weapon could also be of strategic significance in areas of the world without sophisticated nuclear armaments.

However, nearly pure plutonium-239 can be produced in power reactors by removing fuel rods before significant quantities of the other isotopes build up. The fuel would have to be removed and replaced much more frequently than is normal for nuclear power production, a factor which would certainly increase the cost of electricity from the reactor. But that would be unlikely to be an important consideration to a government wishing to produce nuclear weapons, particularly if it wanted only a few. Such frequent refuelling would certainly invite attention if the power reactor were under safeguards. Most types of power reactor have to be closed down for each refuelling, but some types, such as the CANDU, can be refuelled while operating. This would be an advantage if the reactor was being used to produce high grade bomb plutonium.

Historical background

The origins of safeguards against diversion from nuclear power programs to weapons-making, as well as of the IAEA and the Treaty on the Non-Proliferation of Nuclear Weapons (NPT), can be traced back to the period immediately following the Second World War. Demonstration of the enormously destructive power of nuclear weapons had prompted consideration of means of limiting such destructive potential, while at the same time making atomic energy available for the benefit of mankind.

In November 1945, at a meeting of representatives of the Canadian, U.K. and U.S. governments, it was agreed that international action was necessary to prevent the further use of atomic energy in weapons and to ensure the development of its peaceful applications. The three governments declared their willingness 'to proceed with the exchange of fundamental scientific literature for peaceful ends with any nation that will fully reciprocate'. But it was agreed that this exchange could not take place until 'it is possible to devise effective reciprocal and enforceable safeguards acceptable to all nations', which 'would contribute to a constructive solution of the problem of the atomic bomb'.

Discussions followed with the U.S.S.R. Then, in January 1946, the three nations involved in the original discussions and the U.S.S.R., together with France, submitted a draft resolution to the United Nations General Assembly suggesting the establishment of a commission with powers, among others, to make proposals for the control of atomic energy to the extent necessary to ensure its use only for peaceful purposes; for the elimination from national armaments of atomic and all other major weapons of mass destruction; and for effective safeguards, by way of inspection and other means, to protect complying states against the hazards of violations and evasions. At the insistence of the U.S.S.R., the draft resolution made the commission accountable to the U.N. Security Council for those aspects of its work relevant to international peace and security.

This draft resolution was adopted, and the United Nations Atomic Energy Commission was established in January 1946. Five months later, the U.S. submitted to that Commission a comprehensive scheme, known as the Baruch Plan, for complete international control of nuclear energy. This plan proposed an international authority which would own or have managerial control of all nuclear material, from the processing of ore at a uranium mine to its end uses. The power of the authority to impose penalties was an essential element of the proposal. Major violations would have been dealt with by decision of the Security Council, to which the veto power of the permanent members was not to apply. It was also suggested that a control system be set up, after which the stockpile of nuclear weapons in the possession of the U.S.A. (the only nuclear weapon state at that time) would be disposed of.

This plan foundered largely on the inability of the U.S.A. and the U.S.S.R. to agree on the time at which a control system should be established. The U.S. wished the control svstem to be established before the disposal of stockpiled nuclear weapons; the U.S.S.R took the view that the United Nations should first outlaw atomic weapons and require the destruction of such weapons and then proceed to the consideration of a control system. This divergence of views prevented any worthwhile achievements by the U.N. Atomic Energy Commission, which was eventually dissolved in January 1952.

In February 1947, the Security Council established the Commission for Conventional Armaments which, among other things (and with equally little success), considered the international control of atomic energy as part of its plan to regulate and reduce conventional armaments. This body, too, was dissolved in January 1952, in which month the United Nations Disarmament Commission was established.

In December 1953, President Eisenhower made his 'Atoms for Peace' address to the United Nations General Assembly. One of the main objectives of his program to develop peaceful uses of nuclear energy was a start towards diminishing 'the potential destructive power of the world's atomic stockpiles'. To this end, governments principally involved would 'begin now and continue to make contributions from their stockpiles of normal uranium and fissionable materials to an International Atomic Energy Agency'.

It was the U.S. view that these proposals were not matters relating to disarmament. As a result, the three years of negotiations leading to the creation of the IAEA were conducted outside the Disarmament Commission. The main significance of the U.S. attitude was that it represented an abandonment of comprehensive disarmament and its replacement by the search for more immediately practical measures. It was on this latter basis that the IAEA was created in 1957, to foster the spread of atomic energy for peaceful purposes and with defined rights to impose safeguards designed to ensure that nuclear energy was not used in such a way as to further any military purpose.

From the early 1950s, the effective disarmament forum was a subcommittee of the U.N. Disarmament Commission, consisting of Canada, France, the U.K., the U.S.A. and the U.S.S.R. In 1959 its membership was increased to ten by the addition of Bulgaria, Czechoslovakia, Poland, Romania and Italy; and in 1961, with the addition of eight non-aligned powers (Brazil, Burma, Ethiopia, India, Mexico, Nigeria, Sweden and the United Arab Republic), it became the Eighteen Nation Disarmament Committee, which is still in existence.

By 1954, both the U.S.A. and the U.S.S.R. had exploded thermonuclear devices (H-bombs), the U.K. had begun production of nuclear weapons, and France was soon to have significant plutonium production capacity. The problems had become not a matter of nuclear disarmament alone; the additional problem was to limit the spread of nuclear weapons and weapons-making capability.

These are the basic issues with which the Eighteen Nation Disarmament Committee and the United Nations have endeavoured to grapple. In March 1968, the Committee produced a report recommending a draft non-proliferation treaty. Notwithstanding a variety of criticisms, the draft was approved by the General Assembly of the United Nations on 12 June 1968. The NPT was opened for signature on 1 July 1968, and entered into force on 5 March 1970.

The International Atomic Energy Agency (IAEA)

The functions of the IAEA are crucial to any examination of the present safeguards system.

The Agency is a largely autonomous international organisation, which reports annually to the U.N. General Assembly and, when appropriate, to the Security Council. It has a membership of 109 countries, amongst which are all the major nuclear powers, except China. Australia and its principal trading partners are members. The objective of the IAEA is to 'accelerate and enlarge the contribution of atomic energy to peace, health and prosperity throughout the world'. Towards achieving this aim, it was intended initially that the Agency would accumulate nuclear materials for supply to member States; act as an intermediary for securing materials, equipment or facilities; encourage the exchange and training of scientific personnel; and promote the exchange of scientific and technical information. The IAEA is performing these functions, except that it has not become a store-keeper of nuclear materials.

In order to fulfil its role in encouraging the peaceful use of nuclear technology, the IAEA developed two systems of safeguarding nuclear power industries. The initial safeguards were designed to apply to projects involving IAEA assistance, to meet requests by parties to any bilateral or multilateral agreements, and to meet requests by individual countries with nuclear commitments. The more recent safeguards stem from obligations undertaken by states which are party to the NPT.

IAEA safeguards

IAEA safeguards have been shaped by the nature of specific problems and by the degree to which countries will permit their nuclear industries to be regulated. These safeguards normally apply to particular facilities rather than to all facilities in a country. Initial safeguards were developed in 1958 in response to a Japanese request for assistance in obtaining fuel for a heavy-water research reactor. Subsequently, safeguards were expanded in a series of steps to become more generally applicable to a wide range of nuclear processes. The first of these general safeguards systems appeared in 1961 and related to reactors of less than 100 megawatts thermal capacity. In 1965, safeguards were extended to cover reactors with greater than 100 megawatts thermal capacity. This system was revised in 1965, and was extended in 1966 to incorporate provisions relating to reprocessing plants. In 1968 the system was further extended to include provisions for the safeguarding of nuclear material in conversion and fuel fabrication plants. This system, known as 'The Agency's Safeguards System (1965, as provisionally extended in 1966 and 1968)' and reproduced in IAEA document INFCIRC/66/Rev. 2, constituted the extent of safeguards development at the date on which the NPT entered into force.

In February 1974, the IAEA Board of Governors decided that subsequent IAEA safeguards agreements should normally contain provisions which relate the duration of the agreements to the period of actual use of the safeguarded items. The Board also decided that the agreements should normally confirm the right of the IAEA to continue to apply safeguards to produced special fissionable materials until the provisions for termination of safeguards contained in INFCIRC/66/Rev. 2 have been satisfied.

Experience with IAEA safeguards demonstrates that countries have not been prepared to accept continuous surveillance of nuclear activities by an external authority. The control system established by the Agency involves accounting methods augmented by regular 'on the spot' inspections. The inspections are carried out by a team of skilled personnel within the IAEA; as at 30 June 1976, according to information available to the Commission, there were 79 inspectors, of whom about 50 regularly carried out inspections. Several witnesses were of the opinion that this force was far too small to maintain effective surveillance of existing installations covered by IAEA safeguards.

In brief, the accounting procedure is based on a system of records and reports which are maintained by a country with respect to facilities and nuclear materials in its territory. It is then for the inspectorate to carry out regular audits of records and reports, to check the amount of safeguarded material, and to scrutinise the operation of facilities subject to safeguards. Before commencing, the inspectorate carries out a facility design review in order to ensure that effective safeguards can be applied to each facility. Precise details of the safeguards measures to be applied in a particular state are included in a subsidiary arrangement to the safeguards agreement, which is not made public, so as to protect any commercial and industrial secrets of the state.

NPT safeguards

Historically, the NPT originated as a disarmament measure to prevent the spread of nuclear weapons. This is clearly its most significant aspect. Article I of the NPT states that nuclear-weapon countries which are party to the Treaty must undertake not to transfer nuclear weapons or other nuclear explosive devices to any recipient. Article II provides that non-nuclear-weapon countries are not to receive any such transfer, nor are they 'to manufacture or otherwise acquire nuclear weapons or other nuclear explosive devices'. To implement the latter part of this undertaking, each non-nuclear-weapon country has to accept safeguards 'for the exclusive purpose of verification of the fulfilment of its obligations assumed under this Treaty with a view to preventing diversion of nuclear energy from peaceful uses to nuclear weapons or other nuclear explosive devices' (Article III.1). Those safeguards are to be applied by the IAEA.

The NPT requires safeguards to be applied to all 'source' or 'special fissionable material' used in all peaceful nuclear activities within the territory of a country, under its jurisdiction, or carried out anywhere under the control of that country. 'Source material' is defined by the IAEA Statute to mean natural or depleted uranium, or thorium. 'Special fissionable material' means, in effect, enriched uranium or any material containing plutonium-239 or uranium-233. In fact, as described later, safeguards are not applied to source material.

All parties to the NPT are required 'not to provide (a) source or special fissionable material, or (b) equipment or material especially designed or prepared for the processing, use or production of special fissionable material, to any non-nuclear-weapon State for peaceful purposes, unless the source or special fissionable material shall be subject to the safeguards required by this article' (Article III.2).

Safeguards required under the NPT are set out in the IAEA document entitled 'The Structure and Content of Agreements between the Agency and States required in connection with the Treaty on the Non-Proliferation of Nuclear Weapons' (INFCIRC/153). This document constitutes a model agreement.

Although there are many similarities between INFCIRC/153 and the IAEA safeguards under INFCIRC/66/Rev. 2, there are a number of significant differences. Perhaps the most striking is that NPT safeguards are to be applied to all nuclear material in a country, from the stage in the fuel cycle where it is converted to a composition or purity suitable for fuel fabrication or enrichment. (This means that, in contravention of the NPT requirement, source material is not safeguarded.) Moreover, each country is required to establish and maintain a national system of accounting for, and control of, all nuclear material subject to safeguards. Indeed, NPT safeguards emphasize the fundamental importance of material accountancy as a safeguards measure, with containment and surveillance as important supplementary measures. The IAEA is obliged to verify the national accounting system.

NPT safeguards require also that IAEA verification procedures be specially concentrated on those stages in the nuclear fuel cycle which involve the production, reprocessing, use or storage of nuclear material from which nuclear weapons could readily be made. This means that the IAEA should pay more attention to activities at, for instance, a reprocessing plant than to activities at a nuclear reactor operating normally to produce electricity.

NPT safeguards differ also from those under earlier IAEA systems in that they spell out a specific objective for the safeguarding operation, and indicate the nature of conclusions to be drawn from IAEA verification activities and the types of statements which the IAEA should make to a safeguarded country. The objective of the NPT safeguards procedures is stated as being the timely detection of the diversion of significant quantities of nuclear material from peaceful nuclear activities in order to manufacture explosives or for purposes unknown, and the deterrence of such diversion by the risk of early detection.

Another development envisaged in NPT safeguards is the greater use of scientific instruments to measure the quantities of nuclear material in and moving through successive stages of the fuel cycle. These instruments are to be located at key measurement points in the fuel cycle where greater measurement accuracy can be obtained. The physical measurements are made at regular intervals, and enable the amounts of nuclear material in and moving through each of a series of defined areas, termed material balance areas, to be determined. The results obtained are compared with the figures shown by the accounting records. A serious discrepancy between the two sets of figures may be evidence that diversion has occurred.

24647/76—9

Technical devices of various kinds are also used to safeguard the containment of nuclear materials. Thus seals might be affixed to the lid of a reactor to indicate whether it has been opened in the period between normal fuelling operations carried out under Agency supervision. Cameras may also be used for surveillance to indicate whether any attempt has been made to interfere with the seals or, indeed, the reactor lid; they may also be used to effect surveillance of fuel movements into and out of the reactor or to keep watch over material stored in a spent fuel pond.

To enable it to carry out its verification activities, the Agency is required to be supplied by the state concerned with the necessary information relating to nuclear materials and facilities. The results of the verification activities under NPT safeguards are expressed as a formal statement, in respect of each material balance area, of the amount of material unaccounted for over a specified period, and giving the limits of accuracy of the amount stated. The IAEA is required to inform the country concerned of its conclusions as soon as possible after the measurements have been made and verified. The operation of the accounting system and the various ways in which it is checked are discussed in Chapter 13 under the heading **Accounting Procedures.**

Regional safeguards

Regional groupings have contributed to control of nuclear weapon proliferation in Western Europe and Latin America.

The European Atomic Energy Community (Euratom) was set up in 1957 'to contribute to the raising of the standard of living in the Member States and to the development of relations with the other countries by creating the conditions necessary for the speedy establishment and growth of nuclear industries'. In order to perform that task, the Community is required, among other things, to 'make certain, by appropriate supervision, that nuclear materials are not diverted to purposes other than those for which they are intended'. The original member states of the Community were Belgium, the Federal Republic of Germany, France, Italy, Luxembourg and the Netherlands. They have recently been joined by Denmark, Ireland and the United Kingdom.

The principles of Euratom safeguards are similar to those in the IAEA Statute. However, an important difference is that Euratom safeguards do not extend to nuclear materials which are intended to meet defence requirements and are being specially processed or, having been processed, are stored in a military establishment.

In April 1973 an agreement was signed between Belgium, Denmark, the Federal Republic of Germany, Ireland, Italy, Luxembourg, the Netherlands, Euratom and the IAEA in connection with the NPT. This agreement is noteworthy for the degree of co-ordination which it provides between the safeguards activities of Euratom and IAEA. Indeed the IAEA relies to a significant extent on Euratom safeguards in reaching conclusions on diversion.

The other regional grouping is found in the Treaty for the Prohibition of Nuclear Weapons in Latin America (The Tlatelolco Treaty), which came into force in 1967. This Treaty requires parties to refrain from developing, acquiring, testing or using nuclear weapons, and provides for a control system. Under this system, IAEA safeguards apply to all the nuclear activities of each party. In addition to the IAEA safeguarding activities, the General Secretary of the Tlatelolco Secretariat may request special reports from any party regarding any event or circumstance connected with compliance with the Treaty. The Council set up by the Treaty may carry out a special inspection at the request of any party which suspects that a prohibited activity has been or is about to be carried out. It should also be noted that responsibility for detection of the importation of nuclear weapons into the region rests with the Agency established by the Treaty, and not with the IAEA.

Relationship between IAEA and NPT safeguards

Under the IAEA system of safeguards, a state is prohibited from using safeguarded materials or materials from a safeguarded facility in such a way as to further any military purpose. Problems can arise when a state, the facilities of which are subject to safeguards, becomes a party to NPT, because the NPT proscription covers use as weapons or explosives but not military purposes in general. The prime example of military use not prohibited by the NPT is nuclear propulsion systems for naval vessels.

If the state acceding to NPT has obtained materials or a facility under an Agency project agreement, no difficulties will arise. The state's NPT safeguards agreement does have the effect of automatically suspending the application of the Agency safeguards in the project agreement. But, by virtue of a provision in the NPT safeguards agreement (required by INFCIRC/153, paragraph 24) the undertaking in the project agreement not to use nuclear items for military purposes will continue to apply.

The position if the prior safeguards agreement is not a project agreement is less clear. If the prior agreement is between the IAEA and a single country, implementation of the NPT safeguards agreement operates automatically to suspend application of safeguards under the prior agreement. As there is no equivalent saving of the military purposes prohibition, the country may use safeguarded nuclear items for military non-explosive purposes. If the prior safeguards agreement is between the IAEA, a supplier country and a recipient country, then the inclusion of a suspension clause in the recipient country's NPT safeguards agreement cannot operate automatically to suspend the earlier agreement, unless there is a term in the earlier agreement providing that the NPT safeguards agreement is to have that effect. Otherwise, for the suspension to take effect, a further agreement must be signed by the IAEA and the supplier and recipient countries.

Such agreements have, so far, taken one of two forms. In one type, the agreement merely suspends the prior safeguards agreement while the NPT safeguards agreement remains in force; and it would thus appear, on a strict application of that suspension, that the recipient state could use safeguarded items for military non-explosive purposes.

In the other type, in addition to such suspension, a requirement is imposed that the recipient (NPT) state satisfy both the IAEA and the supplier state that the nuclear material to be withdrawn from NPT safeguards is not subject to the peaceful use guarantees made to the supplier state by the recipient state in the Agreement for Co-operation between those states. In a situation where there is such a guarantee, the recipient state may not use items which were covered by the prior safeguards agreement and the co-operation agreement for military non-explosive purposes.

Any such suspension operates, of course, only for as long as the NPT agreement with the IAEA remains in force. Should the NPT safeguards agreement cease to be in force, the prior IAEA safeguards agreement and the prohibition against military use become binding again.

Another major difference between the two safeguards systems is that IAEA controls are normally limited to specific items or projects, whereas the NPT system, in principle, embraces a country's whole nuclear industry, except for military non-explosive uses. In the terminology of the NPT safeguards agreements, it is stated that they 'should provide for the Agency's right and obligation to ensure that safeguards will be applied, in accordance with the terms of the Agreement, on all source or fissionable material in all peaceful nuclear activities within the territory of the State, under its jurisdiction or carried out under its control anywhere'.

Reference has already been made to the provisions of Paragraph 2 of Article III of the NPT relating to the transfer of source or special fissionable material and certain equipment and other material. No difficulty arises on transfer of any of the items mentioned in that Paragraph to a non-nuclear-weapon country which is a party to the NPT, because those items will become subject to NPT safeguards in that country. However, the Paragraph, despite its reference to 'safeguards required by this Article', has not been interpreted as requiring the application of NPT safeguards to items received on transfer from a NPT country by a non-nuclear-weapon country not party to the NPT. The interpretation has been that IAEA safeguards are to be applied to the items transferred. Thus, if a reactor were supplied by a country which is a party to the NPT to a non-nuclear-weapon country not party to the NPT, then IAEA safeguards would be applied to the reactor and any source or special fissionable material processed, used or produced in the reactor.

Conclusions

The IAEA is the major international body responsible for encouraging the development of peaceful uses of nuclear energy and for the implementation of safeguards aimed at inhibiting the use of nuclear energy for non-peaceful purposes.

Several international safeguards systems exist, with various degrees of overlap, conflict and effectiveness. The system established under the NPT is the most widely applied and, in most respects, appears to be the most effective. The weaknesses of these systems are discussed in Chapter 13.

Because enrichment plants are largely confined to existing nuclear-weapon states, proliferation is most likely to occur from fuel reprocessing plants separating out plutonium. Small scale reprocessing plants already exist in Japan, India, Italy, Spain, Yugoslavia, Argentina, Taiwan, and the Federal Republic of Germany, as well as in the nuclear-weapon states.

13 WEAKNESSES OF THE NPT AND OF THE SAFEGUARDS SYSTEM

A wide range of factors was referred to in evidence by witnesses who sought to show that the present safeguards arrangements were so unsatisfactory that little or no reliance could be placed upon them as means of preventing the further widespread proliferation of nuclear weapons.

Membership of NPT

The viability and effectiveness of the NPT and of the international order it is intended to promote are dependent upon the extent to which states adhere to the Treaty and upon the attitudes and significance of those states which do not become parties to it. A total of 105 states have ratified the NPT, and this is an important measure of its likely success. Reference must also be made to a number of the absentees from the list of NPT parties. (Parties and non-parties to the Treaty are listed in Appendix C.)

A major reason given by some countries (France and China, for example) for not joining the NPT is that it is wrong and unsound for a monopoly of nuclear weaponry in the hands of some powers, notably the U.S.A. and U.S.S.R., to be maintained by treaty. Of course, such attitudes are tinged if not governed by practical considerations affecting particular countries.

The existence of a number of states with possible nuclear weapon ambitions has been pointed to by a number of witnesses as a serious flaw in the NPT. Admittedly some states which are not parties to the Treaty have advanced, or comparatively advanced, nuclear capabilities and also grounds for believing that their security interests might require them to be able to make nuclear weapons at short notice. South Africa would be in this category. Israel is in a similar position, but it may already have made atomic bombs. Some states regard themselves as subject to a direct threat from a neighbour: Egypt with respect to Israel's suspected nuclear potential; Pakistan in the light of the Indian 'peaceful nuclear explosion'; Taiwan in relation to China; and Brazil and Argentina with respect to each other. The acquisition of nuclear weapons or explosive devices by any of these states would have the effect of increasing international tensions in their particular region, but the effect on the future of the NPT is less easily estimated. In most cases, the countries which would most obviously be influenced by the acquisition of nuclear weapons by those states refrained from becoming parties to the Treaty because of that very possibility.

Three non-parties, China, India and France, have little in common other than their existing nuclear capacity, but their attitudes will clearly influence the future of the NPT regime. Both China and India have achieved a position of some prestige in the third world as examples of what can be achieved by states that not so very long ago had neither advanced industrial nor technological expertise. India's nuclear explosion was probably, in part at least, a response to the threat which it regards China as posing; while China's was more obviously designed to provide some protection against the possibility of Russian expansion at its expense.

Although not a party, France has repeatedly stated its acceptance of the principles of the Treaty as far as the supply of French nuclear facilities and materials to other states is concerned. The cause of the disquiet about France's attitudes has been its apparent refusal to collaborate with the proposal, now being advanced by the U.S.A. and the U.K. under the auspices of the 'London Suppliers' Club', that an embargo should be placed on sales of 'sensitive' equipment. South Korea abandoned its proposed purchase of a reprocessing plant from France under heavy pressure from the United States, but similar pressure did not prevent arrangements proceeding in a similar transaction between France and Pakistan.

Conflicts in the Treaty objectives

A serious threat to the viability of IAEA and NPT safeguards as a means of restricting nuclear energy to peaceful uses is likely to come from the inherent conflict of aims in the Agency Statute and more particularly in the NPT. Both instruments are based upon the theory that it is possible to segregate the peaceful use of nuclear energy from its weapon-making potential and that, by safeguarding a facility or the materials used in a facility, this theory can be given practical effect. On the strength of this assumption, the Statute is designed to promote the peaceful uses of atomic energy. Article IV of the NPT, as a counter-weight to its undertakings relating to weapons non-proliferation, incorporates an obligation upon all parties to facilitate, and grants to all parties the right to participate in, the 'fullest possible exchange of equipment, materials and scientific and technological information'. We have been advised, and we accept, that this Article does not create a binding legal obligation, and in particular does not bind Australia to mine its uranium and sell it to any particular country, or at all. The fact is, nevertheless, that the NPT only became possible because of the assurances in Article IV concerning the provision of nuclear equipment, materials and information for peaceful purposes.

The Treaty seeks to make difficult the acquisition by non-parties of the necessary equipment, materials and information to develop nuclear weapons or other nuclear explosive devices. Quite obviously, an advanced industrial country not party to the NPT might be able to develop nuclear weapons from its own resources. It could be assisted in one way or another by another nation not restrained by international treaty, whether the NPT or another treaty. The encouragement of peaceful uses of nuclear energy is bound to have some effect in making these avenues available, if for no other reason than because nuclear science and technology will spread. The NPT does not try to do anything about the movement internationally of scientists, engineers and technicians.

The close link between the ability to make nuclear bombs and nuclear power programs has already been discussed, and we will return to the subject later. Progressively, there will be brought about a situation in which a number of countries, perhaps many countries, will be in a position to make atom bombs if they wish to. This must have been foreseen, although the development has probably been faster than anticipated. Viewed in this light, the NPT and the safeguards system generally constitute a massive holding operation. The day of nuclear weapon omnicompetence is delayed, and it is hoped that in the meantime ways and means of handling that situation when it arises will be discovered.

In attaining nuclear competence, India made use of an imported reactor. By that means, it produced material adapted for a device to produce what the Indian government has consistently maintained was a peaceful nuclear explosion. It appears that, in fact, IAEA safeguards did not apply to the reactor in question, nor to any material used in constructing the device. However, it does not follow that this case is thereby rendered of no significance in the present context. It does show that, unless all facilities in a country are subject to NPT safeguards or their equivalent, the availability of an unsafeguarded source of fuel and of the required technology enables a nuclear explosives program to proceed virtually undetected. The IAEA safeguards system only operates as a result of an agreement by which the state or states concerned submit facilities or nuclear material to safeguards. The position is even less satisfactory in the light of the dispute, which we discuss elsewhere, over whether a peaceful nuclear explosive device, which is technically indistinguishable from a nuclear bomb, can be regarded as being for a non-military use within the meaning of the Agency Statute and safeguards agreements.

The Indian explosion has emphasised that, whatever part unsafeguarded facilities or materials might play, the range of nuclear expertise can just as well be obtained from the experience gained in operating safeguarded facilities. This path to a nuclear explosives capacity is probably slower, but it is available to any country which has a commercial nuclear program and the resources, financial and industrial, to back up its intentions. The costs involved in the development of an atomic weapons capability are such that a commercial nuclear program, particularly if it can be designed to include enrichment or reprocessing facilities, or both (on however small a scale), does offer a satisfactory 'half-way-house' to a military objective, and it also provides tangible energy benefits to offset the costs involved.

Use of material for non-explosive military purposes

Another significant weakness in the NPT system, which has been alluded to earlier, is that the treaty only prohibits nuclear weapons and other explosive devices. Unlike the Agency safeguards, NPT safeguards are not designed to prevent the use of nuclear materials for other military purposes (propulsion units being the example usually given); indeed NPT safeguards expressly provide for materials to be withdrawn from the inventory for such a purpose. Thus paragraph 14 of that document, headed Non-Application of Safeguards to Nuclear Material to be Used in Non-Peaceful Activities, provides that when a state intends to exercise its discretion to use nuclear material in this way it should inform the Agency, making it clear (i) that no undertaking exists, in respect of which IAEA safeguards apply, which would prohibit the use of the material in a non-proscribed military activity, and (ii) that, during the period when safeguards are not applied, the material will not be used for the production of nuclear weapons or other nuclear explosive devices.

It is also set down in paragraph 14 that, when a state makes use of this right to withdraw materials from safeguards, the state and the Agency shall make an arrangement for the non-application of safeguards while the materials are in use for non-proscribed military purposes. During this time all that is required is that the arrangement 'shall identify, to the extent possible, the period or circumstances during which safeguards will not be applied' and that the Agency 'shall be kept informed of the total quantity and composition of such unsafeguarded nuclear material in the state and of any exports of such material'.

It need hardly be said that critics of the NPT see in this situation a substantial defect in the reliability of the Treaty's safeguards arrangements.

It must be admitted that there is no means of controlling the use to which the materials might be put once they are withdrawn from controls. Indeed it is highly illogical to have a safeguards system of accountancy and surveillance which is designed to provide some assurance that states are not misusing nuclear materials for weapons-making purposes if any state is able to escape even this limited control mechanism by a mis-statement of its intentions.

Withdrawal from safeguards arrangements

Criticisms have been levelled against existing safeguards arrangements on the ground that the treaties or agreements upon which they are based all include withdrawal clauses entitling parties to terminate their obligations by giving the requisite notice. For example, the Agreement between the IAEA, Japan and Australia for the Application of Agency Safeguards of 28 July 1972 can be terminated 'by any Party upon six months' notice to the other Parties' (section 35). Agreements for the application of Safeguards under the NPT (which are bilaterial arrangements between individual states and the Agency) remain in force for as long as the state concerned is a party to the Treaty. Hence they are dependent upon Article X.1 of that Treaty:

> Each Party shall in exercising its national sovereignty have the right to withdraw from the Treaty if it decides that extraordinary events, related to the subject matter of this Treaty, have jeopardised the supreme interests of its country. *It shall give notice of such withdrawal to all other Parties to the Treaty and to the United Nations Security Council three months in advance*. Such notice shall include a statement of the extraordinary events it regards as having jeopardised its supreme interests. (Emphasis added)

The wide nature of the discretion available to a state that does wish to withdraw is evident. Apart from the purely formal requirements relating to the notice to be given, the only legal fetter upon the exercise of the discretion concerns the limitation that the justification for the withdrawal must be related 'to the subject matter of this Treaty'. It is for the withdrawing party to decide for itself whether its supreme interests are jeopardised. Given the self-adjudging nature of the provision, which takes it outside the scope of review by any international arbitral or judicial body, the scope for withdrawal is extremely wide. Thus, even if international safeguards were in themselves a totally adequate guarantee against diversion by governments, those safeguards are dependent upon treaty arrangements that can be terminated by unilateral act.

This is undoubtedly a serious limitation on the operation of the NPT and of most safeguards arrangements. There are in existence, however, an increasing number of agreements which provide additional or 'back-up' safeguards if the state which has received nuclear materials or facilities does withdraw from NPT and IAEA safeguards. The practice of the U.S.A., dating from before the IAEA was established, is to enter into co-operation agreements which provide for the right of the United States itself to verify that the recipient state has not diverted any materials or equipment supplied under the agreement to a military purpose. Once the Agency safeguards system came into being, the U.S.A. arranged that its own safeguards were to operate only if for some reason the international safeguards were no longer applicable.

'Fall-back' safeguards, to use another term, at least have the effect of providing some means of ensuring that a peaceful purposes undertaking is respected. On the other hand, they do suffer from a number of disadvantages, some of which could be overcome by revised drafting of the agreements.

The Agreement of July 1972 between Australia and Japan for Co-operation in the Peaceful Uses of Atomic Energy provides illustration of the difficulties. Article II (1) states that each state 'shall ensure that the material, the equipment and facilities obtained by it, or by persons under its jurisdiction authorised by it, pursuant to the present Agreement, and special fissionable material recovered or produced as a by-product . . . shall be used only for peaceful purposes'. It is then provided by Article IV that, if the Agency or NPT safeguards referred to in Article III, are 'inoperative', the supplying state should be entitled to exercise a number of rights of examination, inspection, etc. to 'assure itself that the undertakings set out in Article II of the present Agreement are complied with'.

The safeguards only cover items supplied pursuant to the Agreement. An arrangement of this type is now regarded as unsatisfactory. It is being proposed in the United States that future supplies of American materials, facilities, etc., should only be made available to countries prepared to accept back-up safeguards on their entire fuel cycle. This is obviously a desirable step which Australia, despite its less powerful bargaining position, should follow in future co-operation agreements under which uranium sales might take place. Back-up safeguards agreements should not give a right of unilateral withdrawal.

Should Japan ever be in breach of Article II, the Australia-Japan Agreement would not cover plutonium produced from Australian uranium. By Article IX (2), if a party in breach fails to take corrective steps, the other party may terminate the agreement. Existing contracts made under the agreement could be cancelled, and Australia could require the return of any special fissionable material supplied pursuant to the agreement. In the event of a termination of the agreement, enforcement by Australia of such a requirement would seem impracticable. Moreover, the right to require the return of material does not extend to any material, such as plutonium, produced from fissionable material originally supplied. Extension of the requirement in future agreements to cover such products does not seem feasible because of the impracticability of establishing the country of origin of nuclear material once it has been processed.

There would still remain a number of practical difficulties in the application of the 'back-up' safeguards with Japan. The system of records, inventories and inspections operated by the IAEA needs a well-qualified and experienced team to ensure a reasonable level of effectiveness in its application. The U.S.A. and the U.K. would certainly be able to service similar schemes of safeguards in relation to states with which they had 'back-up' arrangements, but the question must be asked whether Australia would be able to call upon sufficient personnel with the expertise to carry out such functions. The imposition of sanctions for breach of international safeguards arrangements is the subject of specific provision in the Agency Statute, which places responsibility on the U.N. Security Council and General Assembly as well as on the Agency itself.

Although a supplier state would be entitled to report breach of any treaty, and therefore of back-up safeguards arrangements, to the political organs of the United Nations, it would be reasonable to suppose that the Security Council or General Assembly would feel less disposition to take some form of action in a case of breach of a bilateral agreement between states than in the case of breach of an international arrangement involving the Agency. It would follow, therefore, that the efficacy of back-up safeguards might well depend upon the degree of pressure that the supplier state, or states, would be able to exert against the particular diverter state. Here again, the United States, for example, would be better placed to secure compliance with safeguards requirements than would a country like Australia. There is much to be said, therefore, for the proposition that, if Australia is seeking new markets for uranium, it should limit its supplies to countries which are subject to back-up safeguards obligations with states, such as the United States, better able to exert the necessary pressure to ensure compliance.

Further transfer of materials

The NPT does not prohibit the further transfer of materials by a receiving state to a third state, and is not entirely satisfactory in the provision it makes for safeguards on such retransfers.

By Article III.2 of the NPT, each party to the Treaty 'undertakes not to provide (a) source or special fissionable material, or (b) equipment or material especially designed or prepared for the processing, use or production of special fissionable material, to any non-nuclear-weapon State for peaceful purposes, unless the source or special fissionable material shall be subject to the safeguards required by this Article.' There are a number of ambiguities in this provision. The most significant is that, on the face of it, the 'safeguards required by this Article' should refer to the comprehensive NPT safeguards applicable to the entire nuclear industry in a state (i.e. all its facilities and material, and not just the item actually transferred to the non-NPT state). However, what has occurred in practice is that the view has been taken that, on transfer of designated materials or equipment to a non-nuclear-weapon state not party to the Treaty, all that is required is the application of IAEA safeguards to the transferred items.

The position is not entirely satisfactory if the retransfer is to an NPT party. In such a situation there is no need for the transferring state to impose any restrictions, because the materials or equipment will automatically fall under NPT safeguards. However, it has already been pointed out that, because any state can withdraw from NPT, a supplier state should insist upon some back-up agreement for the continuation of safeguards to meet such an event. In the case of retransfer, there is no obligation under NPT for the retransferring state to require additional safeguards as a precaution against the state receiving the materials or equipment withdrawing from that Treaty. Hence, if the original supplier state wishes to ensure that its materials or equipment do not become free of safeguards in this fashion, it must insist upon some provision in the agreement with the first recipient state which will ensure that any retransfer will only take place subject to adequate safeguards.

There are other problems concerning application of adequate safeguards on retransfer. The safeguards required by Article III of the NPT are set out in INFCIRC/153. Paragraph 34 of that document distinguishes between, on the one hand, 'nuclear material of a composition and purity suitable for fuel fabrication or for being isotopically enriched', and, on the other hand. 'any material containing uranium or thorium which has not reached' that stage of the nuclear fuel cycle. On a transfer to a non-nuclear-weapon state, in the case of the former category of material, full NPT safeguards apply; but, in the case of the latter category of material, there exists only an obligation to inform the IAEA of the quantity, composition and destination of that material. Hence, if Australia, being a party to the NPT, were to sell a quantity of yellowcake to a non-nuclear-weapon state also party to the Treaty, such as Japan, there would be an obligation on each to report the transfer to the IAEA, but no accounting or other safeguards would be imposed.

The NPT does not prevent Japan from reselling that material to another state. Should Japan choose to resell the yellowcake to an NPT party, the only obligation binding Japan and the purchasing state would be to notify the IAEA of the transaction. If, however, Japan should choose to retransfer the yellowcake to a non-nuclear-weapon state that is not a party to the NPT, Japan, in accordance with Article III.2 of the NPT, would be obliged to require the imposition of IAEA safeguards; and yellowcake would then be subject to full safeguards and not merely to a reporting requirement.

Concern has been expressed that NPT safeguards only require notification to the IAEA of transfers of yellowcake. A state can, in this way, acquire quantities of material which are not subject to any significant controls, and which may be diverted to weapons production. The Commission regards this as an unfortunate weakness in NPT safeguards, but the principal defect lies in the fact already alluded to: that it is possible for any state to withdraw from the Treaty or agreements upon which the safeguards arrangements are based.

The simplest means of dealing with the three situations (i.e., where the transfer is to a non-NPT state, where it is to a party to that Treaty, or where the material being transferred is yellowcake) is to require, by a provision in a co-operation agreement containing the fall-back safeguards, that no retransfer is to take place without the consent of the original supplier (i.e., Australia, in our example). In this way, not only would Australia be able to oppose its materials and equipment passing to a state it would not itself wish to provide with nuclear items, but it would also be in a position to refuse to approve any retransfer until satisfactory safeguards arrangements were made. Although it would be possible to include provisions regulating retransfer in the commercial contract relating to the original sale, it would be unsatisfactory to rely upon the efficacy of such provisions alone, because there could well be problems in seeking to enforce them in the courts of foreign states.

Accounting procedures

As has already been stated, material accountancy is of fundamental importance in the NPT safeguards system. The IAEA pays particular attention to those stages of the nuclear fuel cycle where material from which nuclear weapons could be made is most vulnerable to possible diversion—reprocessing plants rather than reactors, for example. In carrying out this work the Agency also aims to concentrate its efforts at large nuclear fuel cycle facilities where, because of the large amounts of nuclear material being handled, the problems of detecting diversion of a given amount are greatest. The IAEA also categorises material according to its importance for safeguards, so that, for example, more effort would be devoted to safeguarding plutonium than to low enriched uranium.

The essence of the accounting part of the safeguards system consists of two forms of independent checks carried out by the IAEA on the existing national accounting system. As an illustration of how the accounting system works, we confine ourselves to a discussion of plutonium because it has been central to the fears expressed by a number of witnesses.

Plutonium is produced in a reactor. If and when plutonium recycling is established, it will be extracted from the spent fuel rods at a reprocessing plant, transported to a fuel fabrication plant to be made up into fuel rods, and then, contained in the rods, transported to another reactor. The cycle is divided into a number of material balance areas (MBAs) which are chosen so as to enable physical measurements to be made of the amount of plutonium contained within and moving into or out of the area. From knowledge of the design and operation of all parts of this cycle it is possible to make up accounts showing the amount of plutonium present in or moving through each material balance area. A material balance area would normally consist of a single plant, e.g., a reprocessing plant, but there might be more than one in a plant if it were very large. There may be many MBAs in a country with a large nuclear industry. The records for each material balance area are sent to the IAEA inspectorate, who compare them with their own records and check them thoroughly for errors and inconsistencies. In this way it is hoped that diversion based on account falsification would be detected. This constitutes the first check on the national accounts.

As part of the national accounting system itself, the records of amounts of plutonium in and passing through each material balance area are compared with the results of physical analyses to determine the actual quantities of plutonium. Allowance is made for losses of material deposited on equipment, dust losses and so on. This comparison is done at regular intervals of a few months, the interval depending on the size of the plant and other factors. It is usually found that the results of the physical analyses are not identical with the figures in the accounting records because of uncertainty of measurement and estimation of the quantity of material deposited on equipment, dust losses and so on. The difference between the two is termed Material Unaccounted For (MUF). If no diversions or other unexpected losses have occurred, MUF should ideally be zero. However, because of the various uncertainties, it usually will not be. The value of MUF actually obtained can be regarded as an approximation, subject to error, of the quantity actually lost. The greater the difference between the value of MUF obtained and zero, the more likely it is that a loss or diversion has in fact occurred. From knowledge of the operation of the plant, the capabilities of the measuring instruments and other factors, it is possible, using statistical theory, to estimate the probable error associated with the result of the physical analyses, and hence with MUF.

These results are shown to the IAEA inspectors, who make their own independent physical analyses as a second check on the national system. They are then able to define a quantity of plutonium such that, if that amount were diverted, the Inspectorate would have a 95 per cent probability of finding a significant discrepancy on the first occasion that it happened. This quantity is sometimes called the Goal Quantity. In such an event, it would not be possible, because of the statistical uncertainties, to be sure of the quantity which had been diverted, or in fact that any diversion had occurred at all. However, the detection of a significant discrepancy would result in further measurements and examination of records with the intention of discovering what had occurred.

The errors associated with MUF are related to the total quantity of plutonium (the throughput) passing through the plant during the period concerned and are commonly expressed as a proportion of that throughput. The Goal Quantity is also expressed in this way. For plutonium in a reprocessing plant, the IAEA has found this to be as high at 5 per cent of throughput; however there is evidence that in some plants the proportion would be smaller. It is less for other stages in the fuel cycle, where more accurate measurements are possible, such as in uranium enrichment. Continuing technical improvements in measuring equipment and analytical techniques can be expected to reduce the error associated with MUF.

In setting up its own accounting system to check a national accounting system, the IAEA compares the Goal Quantity with the amount of plutonium needed to make a nuclear weapon. For a large reprocessing plant, with present levels of measurement accuracy, this amount is less than the Goal Quantity, i.e., the probability of detecting a single diversion of enough plutonium to make a nuclear weapon would be less than 95 per cent. Some improvements to this situation could be made by decreasing the size of material balance areas and increasing the frequency of measurements. However, there is a limit to what can be achieved by such means because of the number of analyses that must be made, the requirement for skilled personnel, and the interruptions to the commercial operation of the plant that result from an extensive measurement effort. Other safeguards measures, such as containment and surveillance, are therefore essential.

Witnesses referred to a quantity called Limit of Error of Material Unaccounted For (LEMUF), which is a statistical measure of the uncertainty in MUF (i.e., the standard error) and is used to estimate the Goal Quantity. For plutonium in a reprocessing plant, LEMUF is about one per cent of throughput. It was repeatedly argued that this figure represents the amount of plutonium that could be diverted from a material balance area during each measurement period. However, it follows from the definitions of MUF and LEMUF that such a view is based on a misconception of the statistical tests involved, and that there is no simple relationship between LEMUF and the quantity which might be diverted without detection.

It can be said that a diversion as large as 5 per cent of throughput—the Goal Quantity—would almost certainly be detected. Indeed, it follows from the definition of goal quantity that the probability of detection would be 95 per cent. The probability of detection decreases as the quantity diverted is reduced and, with amounts very much less than 5 per cent, the diverter would have a good chance of avoiding detection. However, if only a small amount of plutonium was diverted during each measurement period, it would be necessary to make repeated diversions in order to accumulate enough plutonium to form the basis of a nuclear weapons program. Repeated small diversions would be likely to show up as a consistent bias in the values of MUF, which could be revealed by statistical analysis. Nevertheless, such a bias could also be attributed to underestimation of accidental material loss.

If a government, in building a new reprocessing plant or modifying an existing one, chose to use a new process, not previously employed in a reprocessing plant, the situation might be slightly different. Although

Plate 9. This nuclear complex in France contains, among other installations, two thermal nuclear power stations (right middle ground), a reprocessing plant (centre middle ground), the Phenix 250 megawatt prototype fast reactor (beside the river) and a pilot plant for solidifying nuclear waste by vitrification. Called the Marcoule Production Centre, it is located in the Rhone Valley in southern France. (Photo by courtesy of Commissariat a l'Energie Atomique.)

the IAEA examines the design, construction and operation of all new or modified plants in setting up its system of checks, in the case of a new or modified process, lacking long experience with the operation of similar processes, the IAEA inspectors could not be completely sure of what range of errors to expect in the physical measurements. In such circumstances the Inspectors would be likely to look for diversion on finding values of MUF less than the Goal Quantity, i.e., when the probability that diversion had occurred was less than 95 per cent. Even so, it might be possible for a country using such a process change to hide repeated diversions of small amounts of plutonium. Such an approach to diversion could only be attempted by a government having a high degree of technical competence at its command, because of the difficulty of introducing new technology into a reprocessing plant.

Some more general problems with the accounting procedures are also apparent. For example, given the present levels of measurement error, it is difficult to believe that regular diversion of a very small amount (say 0.5 per cent) would result in a bias in the MUF sufficient to be detected by accounting procedures. In a country with a large nuclear industry and many MBAs, this could amount to a considerable quantity.

There is also the possibility that the operating state might regularly produce slightly distorted figures to hide the consistent diversion of small amounts. Whether such diversions would be detected by the IAEA inspectors is a matter upon which the Commission has received insufficient evidence.

Problems arise with the time taken to confirm diversion. If relatively small diversions are occurring, a considerable period might elapse before the initial detection of a value of MUF larger than the Goal Quantity. This problem would be accentuated by any delay in analysis by the IAEA of materials accounts, said to be considerable at present. Repeated measurements and analyses would be essential before the occurrence of diversion could be confirmed. Despite the best efforts of the Inspectorate, if a state had diverted (rather than an employee on behalf of a non-government organisation), obfuscation by such measures as loss of records, breakdown of plant, spillage of materials, hindrance of IAEA inspectors, and claims that the Inspectorate had erred could be expected to prolong the delay before the Inspectorate could feel sufficiently confident to notify the Board of Governors formally that a diversion could not be ruled out. Further delays would be probable before the Board reported to the U.N. and to member states of the IAEA. Subsequent action by the U.N., if any, might not be prompt. The Commission has been told that a nuclear bomb could be made within ten days of acquiring the nuclear material. It is concluded that a country with a large reprocessing industry could make many nuclear weapons before being formally detected by accountancy procedures. Manufacture of a few such weapons would appear feasible even if the reprocessing plant were small.

However, we wish to point out that the Agency itself recognises that accounting procedures alone may not be entirely adequate. Hence, as has already been mentioned, reactors are being protected by seals and by electronic surveillance devices. The Agency acknowledges that, should sensitive equipment (enrichment facilities or reprocessing plant) be supplied more widely, as is proposed in relation to Brazil and Pakistan, the potential for diversion, particularly if full fuel cycle safeguards are not being applied, will be increased. It has been suggested that control over such plants would need to be exercised by some form of continuous surveillance. There is insufficient evidence to allow the assessment of the probable effectiveness of such surveillance.

Financial burden of safeguards

The Commission recognises that, if one accepts even the more conservative estimates of the development of peaceful nuclear energy programs in a large number of countries during the rest of this century, the costs of the application of safeguards by the IAEA might increase to an extent where there would be opposition by member states to paying such a substantial bill.

The extent of any increase in costs will depend on a number of factors. If safeguards procedures were retained in their present form, many more inspectors would have to be employed by the IAEA to safeguard effectively the increased amounts of materials and numbers of facilities requiring safeguarding. However, it is recognised that significant savings could be made if tasks now carried out by the inspectorate were minimised or replaced by developments in safeguards instrumentation.

Under present arrangements, all members of the IAEA contribute to the financing of safeguards as part of their annual contributions to the Agency budget, which are based on the scale of contributions to the U.N.

Some dissatisfaction with this method of financing has already been expressed by a number of states with few or no nuclear activities. They feel that each state should contribute to financing safeguards in proportion to the size and importance of its nuclear activities. This approach would result in the United States paying fifty to sixty per cent of the cost of safeguards. One variation in the application of this principle would be to require each state commercially involved in supplying nuclear materials, equipment or facilities to set aside a proportion of its profits from such transactions to pay for safeguards.

Even if the problem of covering the costs of increased safeguards can be met satisfactorily, there will almost certainly be difficulties in obtaining suitable experienced personnel to carry out the inspection duties. There is only a limited reservoir of such talent, and most of this is in countries with existing nuclear energy programs. Most international organisations—and the IAEA is no exception in this regard—require staff appointments to be made from as wide a spread of countries as possible (the principle of equitable geographical representation, as it is called). Trained personnel are a resource which developing countries can ill-afford to spare, even less so in this very specialised field.

Security issues and the attitudes of nuclear-weapon states

If states are to forego the development of nuclear weapons, they might want some form of undertaking that, should a threat ever be made against them by a state that has developed nuclear weapons clandestinely or otherwise, they can rely upon the protection of one or more of the major nuclear-weapon states. In addition, these states would wish to be assured that existing nuclear-weapon states would not seek to take advantage of their possession of those weapons to threaten the security of non-nuclear-weapon states.

In the week following the commendation of the final text of the NPT by the General Assembly of the U.N., in June 1968, the Security Council adopted Resolution 255. This instrument expressly *took into consideration* the concern of certain states that, in conjunction with their adherence to the NPT, 'appropriate measures be taken to safeguard their security'. However, the operative part of the resolution no more than *recognised* that 'aggression with nuclear weapons or the threat of such aggression against a non-nuclear-weapon state would create a situation in which the Security Council, and above all its nuclear-weapon state permanent members, would have to act immediately in accordance with their obligations under the United Nations Charter'; *welcomed* 'the intention expressed by certain states that they will provide or support immediate assistance, in accordance with the Charter, to any non-nuclear-weapon State Party to the Treaty on the Non-Proliferation of Nuclear Weapons that is a victim of an act or an object of a threat of aggression in which nuclear weapons are used'; and *reaffirmed* the inherent right of individual or collective self-defence, which existed in any case by virtue of Article 51 of the Charter.

This resolution falls short of satisfying the needs of non-nuclear-weapon states in a number of ways.

In the first place, it does not create any obligation upon members of the Security Council additional to that contained in Article 39 of the U.N. Charter, under which it is for the Council to 'determine the existence of any threat to the peace, breach of the peace, or act of aggression' and to 'make recommendations, or decide what measures shall be taken . . . to maintain or restore international peace and security'.

Secondly, Resolution 255 provides no undertaking by the nuclear-weapon states not to use or threaten to use those weapons against a non-nuclear-weapon state. Nor does it offer any guarantee that a permanent member of the Security Council which acted in this way would be unable to use its veto to prevent effective action being taken by the Council under Article 39.

At the time of the adoption of Resolution 255, the United States, the Soviet Union, and the United Kingdom each made a formal declaration to the Council amplifying its attitude to this issue. The Americans, for example, acknowledged that aggression with nuclear weapons, or the threat of such aggression, against a non-nuclear-weapon state 'could create a qualitatively new situation in which the nuclear-weapon states which are permanent members of the United Nations Security Council would have to act immediately through the Security Council to take the necessary measures to counter such aggression or to remove the threat of aggression'; and also affirmed the intention of the United States 'to seek immediate Security Council action to provide assistance, in accordance with the Charter, to any non-nuclear-weapon state' party to the NPT 'that is a victim of an act of aggression or an object of a threat of aggression in which nuclear weapons are used'. However, the permanent members of the Council were not prepared to give an undertaking that they would not themselves use nuclear weapons or threaten to use them against non-nuclear-weapon states party to the NPT.

This issue cannot be divorced from the implications of Article VI of the NPT by which each party 'undertakes to pursue negotiations in good faith on effective measures relating to cessation of the nuclear arms race at an early date and to nuclear disarmament, and on a Treaty on general and complete disarmament under strict and effective international control'. Even if one regards the references to disarmament as pious hopes rather than practical objectives, it nevertheless remains as a reasonable expectation that a renunciation of nuclear weapons development by non-nuclear-weapon states (thus preventing further horizontal proliferation), should be balanced by a cessation of the nuclear arms race amongst the nuclear-weapon states (thus preventing further vertical proliferation).

So far, the principal arrangements that could be considered as implementing Article VI of the NPT are the agreements resulting from the Strategic Arms Limitation Talks of 1972 and 1974 (SALT I and SALT II) between the U.S. and U.S.S.R. The principal outcome of SALT I was a Treaty (the ABM Treaty) on the Limitation of Anti-Ballistic Missile Systems to two on each side (subsequently reduced by agreement in 1974 to one on each side). In addition, an Interim Agreement was reached on Certain Measures with respect to the Limitation of Strategic Arms, which included undertakings not to start construction of additional fixed land-based intercontinental ballistic missile (ICBM) launchers after 1 July 1972; not to convert land-based launchers for light or older type ICBMs to more modern heavy ICBMs; and to limit the submarine-launched ballistic missile launchers and submarines in accordance with the provisions of the Protocol to the Treaty.

24647/76—**10**

There were a number of obvious defects in the arrangements. For example, the ABM Treaty is subject to a withdrawal clause essentially the same as that contained in the NPT, while the Interim Agreement was only intended to operate for five years and to be replaced by 'an agreement on more complete measures limiting strategic offensive arms' (Article VIII.2). Nevertheless, the agreements were greeted at the time as a hopeful sign of a new detente between East and West.

Viewed from that standpoint, the results of SALT II were a major disappointment. At a Summit meeting in Moscow in July 1974, between President Nixon and Mr Brezhnev, agreement was reached on an upper limit on the size of underground nuclear test explosions of 150 kilotons. And, at Vladivostok, in November of that year, it was agreed in principle by President Ford and Mr Brezhnev to limit until 1985 the number of strategic bombers and missile launchers (whether sea or land based) to 2400 on each side, and the number of multiple independently targetted re-entry vehicles (whether sea or land based) to 1320 for each side. Since then little more has been achieved, although in May 1976 the treaty of July 1974 limiting the size of underground nuclear test explosions received some amplification in a Treaty between the U.S.A. and the U.S.S.R. on underground explosives for peaceful purposes and in a Protocol to that Treaty.

Because SALT II was thought to achieve so little, the combined effect of the two sets of agreements has done little to suggest that a halt has been called to the nuclear arms race. Indeed, as measures to reduce vertical proliferation and thus provide an incentive to prevent horizontal proliferation, they have too many defects. It has been pointed out that they have not required the abandonment of any program to which the two states were already committed (the limit on anti-ballistic missile systems being dictated by economic rather than by strategic factors). Nor do they place any restriction on research into the development of new weapons systems. Moreover, it has been suggested that the limits which they do impose are too high to be of any value.

While no one denies that a substantial reduction in nuclear armaments will only be possible on the basis of agreement between the United States and the Soviet Union, the prospects for the non-proliferation of nuclear weapons are also influenced by the attitudes and policies of the lesser nuclear-weapon states. The United Kingdom has insisted upon retaining its own independent nuclear deterrent. Neither France nor China has become a party to the NPT, and both have expressed similar objections to the underlying philosophy of the Treaty.

When France decided to pursue the development of its *force de frappe,* it expressed the view that it was dangerous that nuclear weapons should be the sole prerogative of two or three states and that additional nuclear-weapon states would have a restraining influence on the two 'superpowers'. This argument is basically not dissimilar from the Chinese attitude which regards the NPT as a means whereby the Soviet Union and the United States are seeking to preserve their freedom to pursue imperialistic policies towards non-nuclear-weapon states. To this underlying principle the Chinese have added the qualification that, unless there is general nuclear disarmament, nuclear proliferation whereby socialist, and not imperialist, states obtain nuclear weapons would assist in the preservation of peace. There is thus a variety of strategic factors which might tend to undermine the NPT in the longer term.

It is perhaps too easy, looking at the international scene from an Australian point of view, to play down the attitudes of many countries towards their security interests. The NPT has received a ready acceptance among states that have reason to believe that, because of their position under the nuclear umbrella of the Soviet Union or of the United States, they are relatively immune to nuclear threats from other states. Australian defence policies are ultimately based upon the proposition that, in the last resort, Australia could look for assistance to the United States. Far from this being true for many states in Africa, Asia and the Middle East, there is a genuine fear that the United States might actively interfere upon their territory. Indeed, there has been discussion by some people in the United States since 1973 of the possibility of using force to prevent any future curtailment of necessary supplies of raw materials, particularly of oil, to American industry.

Sanctions

A number of witnesses expressed the opinion that the NPT is inadequate because of the absence of any effective system of sanctions for its enforcement.

This view is an over-simplification and, in the evidence, appears to have become confused with the question of withdrawal, which has already been discussed. The NPT, like a number of other disarmament or arms limitation agreements, is essentially a political declaration in the sense that it, like the others, can be denounced by a state which decides that its supreme interests have been jeopardised. However, as long as the NPT remains in force for a state, it subjects that state to a series of patently legal obligations set out in the Treaty and amplified in the safeguards agreement with the IAEA.

Whenever it negotiates a safeguards agreement, whether of an IAEA or NPT type, it is the IAEA's policy to insist upon a provision entitling it to employ its powers under its Statute in case of a breach of the agreement. Those powers are found in Article XII. Under this Article, it is for the Agency's inspectors in the first place to determine whether there has been compliance with the undertakings contained in safeguards agreements. The inspectors must then report any non-compliance to the Director-General 'who shall thereupon transmit the report to the Board of Governors'.

It would seem, from the wording of Article XII.C, that the Board's first task is to examine the report from the inspectors. A period of discussion would then ensue with the state concerned. If the Board makes a formal determination that there has been a case of non-compliance, it is required to take two steps: to call upon the recipient state to remedy the non-compliance and to report the non-compliance to all members of the IAEA and to the Security Council and General Assembly of the U.N. Following out the procedures might well occupy a lengthy period of time, during which the offending country might be proceeding with the manufacture, or even the use, of nuclear weapons.

Should there be a failure by the recipient state 'to take fully corrective action within a reasonable time', the Board is entitled to 'direct curtailment or suspension of assistance being provided by the Agency or by a member, and call for the return of materials and equipment made available to the recipient member' (a possibility also covered by Article XII.A.7). In addition, the Agency may, 'in accordance with Article XIX, suspend any non-complying member from the exercise of the privileges and rights of membership'. Under para. B. of Article XIX, a member 'which has persistently violated the provisions of this Statute or of any agreement entered into by it pursuant to this Statute may be suspended from the exercise of the privileges and rights of membership by a General Conference acting by a two-thirds majority of the members present and voting upon recommendations by the Board of Governors'.

The political organs of the U.N. are appropriate bodies to take executive action if there is a breach of an international obligation. This is recognised by Article XII.C. of the Statute, referred to above, in the context of non-compliance with safeguards arrangements. Moreover, the diversion of nuclear materials to military purposes contrary to the Agency Statute, or for making weapons or some other explosive device contrary to the NPT, is patently a matter closely related to international peace and security and therefore within the purview of the U.N. Security Council and General Assembly by virtue of the provisions of the U.N. Charter.

The possibility that a dispute might arise as to whether there has been a breach of the agreement between the IAEA and the state or states party to a safeguards agreement has been foreseen. It is the invariable practice to include in each safeguards agreement a clause providing for the creation of an arbitral tribunal and bestowing jurisdiction upon that tribunal in respect of any disputes arising out of the agreement. The decision of this tribunal is binding on the parties to the dispute. Under INFCIRC/153, the provision for arbitration is specifically inapplicable to the diversion of nuclear material. If the Board of Governors is unable to verify that there has been no diversion, it may take direct action, as provided for in the agreement, without resolution of any dispute. This does not apply to agreements under INFCIRC/66/Rev. 2; however these do recognise that the Board of Governors may require that its decisions be implemented immediately, pending the final settlement of any dispute.

The existence of disputes clauses of this type is not significant where a state decides to withdraw from the agreement. Withdrawal, providing it takes place in accordance with the terms of the agreement, not only has the effect of terminating from that moment all obligations under that agreement, but it also brings to an end any obligation to submit to arbitration even in respect of disputes arising before the moment of withdrawal. Even if a decision were given in respect of a particular dispute, there remains the problem of securing compliance with the tribunal's award. In other words, the successful litigant state might still be dependent upon other means, including the support of the political organs of the U.N., to enforce the judgment.

It is beyond the scope of this Inquiry to examine in depth the efficacy of community methods, through the U.N., of enforcing international obligations in general. However, it is possible to draw inferences from current attitudes.

Attention has been drawn to the failure of U.N. organs to act as effective instruments in settling disputes or in punishing allegedly wrongful conduct by individual states. In one sense this criticism is valid. But it must be pointed out that the failings are due not so much to the U.N. as such, or to its structure, as to the fact that the U.N. can only reflect the conflicts of ideologies and interests between its member states. If there is an adequate consensus among member states for a particular policy, the U.N. as an organisation can take effective action in furtherance of that policy provided the right of veto is not exercised in the Security Council. In most situations that consensus is missing, at least to the extent that a substantial minority (in numbers or influence) can prevent a policy being adopted. In view of the security implications likely to arise in cases of governmental diversion of nuclear material for proscribed purposes, it is doubtful whether effective community responses through the U.N. can be expected.

If there is no effective response from the Security Council, what can a supplier state do in a situation of non-compliance with the Board's directive; and what further steps can it take to prevent the wrongdoer from obtaining significant benefits from its wrongful act, whether or not it returns any materials and equipment?

Sanctions for breach of an international obligation are not necessarily channelled through the U.N. Reference has already been made to the measures which can be taken by the IAEA, in addition to reporting non-compliance to the U.N.

In addition to action taken through international organisations, purely economic measures, such as a complete prohibition on future transfers of nuclear materials, equipment, etc., to the state concerned, would be dependent upon the support of suppliers of those items. It is possible that the London Suppliers' Club might encourage the creation of policies to overcome national commercial considerations in the interests of presenting a common front against states seeking to divert nuclear materials in breach of safeguards agreements. However, even if a common approach could be engendered, the outcome would probably only be to hinder, but not prevent, development of the weapons-making capacity of the recipient state.

Unless the supplier state and its associates are prepared to resort to force to close down the facility or facilities used in the activities involved in the breach, and perhaps to neutralise or remove all fissionable materials produced in any such facilities, the only possibility for trying to secure compliance with safeguards requirements would be to mount an extensive economic campaign. Here, again, the extent of the support for such a policy is crucial.

It is often suggested that international law is a weak system of law because of the lack of adequate sanctions to enforce its rules. In most areas of international relations, states are willing to rely upon the reciprocal interest that they each have in abiding by the obligations of customary international law or laid down in treaties to which they are parties. When the security of a state is involved, however, it is less willing to place its vital interests in jeopardy by adhering to a permanent prescription of how it should conduct its policies.

The absence of satisfactory enforcement procedures is undoubtedly a serious flaw in the NPT system, and this is reflected in the largely unrestricted right of withdrawal available as a last resort to parties to the Treaty. It would add considerably to the potential durability of the Treaty if the principal states supplying nuclear materials, facilities and technology could agree upon measures likely to deter states from developing clandestine nuclear weapons.

Peaceful nuclear explosions

Nuclear explosive devices might seem to have a number of possible uses for peaceful purposes. Among those suggested have been the breaking up of rock formations to release inaccessible natural resources such as gas or minerals and the creation of storage sites for gas (underground) or for water (on the surface). In addition, there have been proposals that nuclear explosions could be used for blasting away earth and rocks to divert a river or to create a canal or harbour.

There have been some peaceful explosions in the Soviet Union and in the United States. Although the Soviet Union is still pursuing a number of ventures, American interest in similar projects has waned. The reason for this is partly that of cost, and partly environmental objections to the use of nuclear energy in such a way. More particularly, in the case of surface 'excavations' there are serious unsolved problems of radioactive fall-out and contamination.

The possibility that nuclear explosive devices might be used for peaceful purposes has been a matter of concern in the context of the non-proliferation issue for two reasons. In the first place, there is the question of whether a state could legally justify the diversion of nuclear materials subject to safeguards by arguing that the device being manufactured was for some peaceful use only. Secondly, to the extent that international arrangements do proscribe the development of 'peaceful' nuclear explosives by non-nuclear weapon states, there is the question of how such a state might be allowed the use of an explosive device supplied by a nuclear weapon state under controls adequate to prevent the former state from obtaining technical information that would enable it to advance a nuclear weapons program.

As far as the first issue is concerned, the Indian 'peaceful nuclear explosion' of 1974 brought into the open a controversy which already existed over the distinction between such an explosion and a nuclear bomb. The attitude of the Indian government was that no safeguarded material had been used in the preparation of the device, and even if IAEA safeguards had applied, India would not have been in breach because those safeguards only prohibited the diversion of materials for military purposes. The Canadian and American reaction to this latter proposition was that, as existing technology could not differentiate between a peaceful nuclear explosive device and a nuclear weapon, Agency safeguards prohibited their use. Safeguards agreements negotiated by the IAEA since 1974 have invariably included a clause prohibiting the use of any safeguarded items for, amongst other things, the manufacture of any nuclear weapon or any other nuclear explosive device.

The same difference of opinion has been expressed in relation to the Tlatelolco Treaty. Under the Treaty, Article 1 lays down a total proscription on the testing, use, production, acquisition, receipt, storage, deployment, etc., of nuclear weapons by the parties or by anyone on their behalf in order to establish what the preamble refers to as 'the military denuclearisation of Latin America'. At the same time, however, it was agreed in Article 18 that the Parties may 'carry out explosions of nuclear devices for peaceful purposes—including explosions which involve devices similar to those used in nuclear weapons—or collaborate with third parties for the same purpose'. Not only does this Article expressly recognise the difficulty, or even the impossibility, of distinguishing between the two types of explosive device, but the definition of a nuclear weapon in Article 5 as 'any device which is capable of releasing nuclear energy in an uncontrolled manner and has a group of characteristics that are appropriate for use for warlike purposes' makes no attempt to resolve the dilemma.

In signing Additional Protocol II to the Tlatelolco Treaty, which comprises certain undertakings by nuclear-weapon states to respect the principle of denuclearisation and to refrain from using or threatening to use nuclear weapons against parties to the Treaty, both the United Kingdom and the United States made separate statements on their interpretation of Article 18. The United States, for example, reiterated the fact 'that the technology of making nuclear explosive devices for peaceful purposes is indistinguishable from the technology of making nuclear weapons'. It followed therefore 'that Articles 1 and 5 restrict accordingly the activities of the contracting parties under paragraph 1 of Article 18'. On the other hand, amongst states signing the text of the Treaty, both Argentina and Brazil have stated their belief, from which Mexico expressly dissented, that the development of peaceful nuclear explosions is at present allowed by the Treaty.

The extent to which a state might justify the diversion of materials subject to IAEA safeguards to the development of a weapons-making project by reference to a claim that the devices in question will be put to peaceful use is obviously a problem. States have an undoubted tendency to construe ambiguous treaty provisions in a manner most favourable to the policies they wish to pursue. As long as the change in wording of safeguards agreements approved by the Agency since 1974 is followed, the problem should be restricted to safeguards agreements existing before 1974. The problem will also be of less importance because an increasing proportion of safeguards arrangements will be under the NPT. The proscription in Articles I and II of the NPT relate to the manufacture or acquisition of 'nuclear weapons or other nuclear explosive devices'.

At the time the NPT was drafted, there already existed a genuine interest in the potential of nuclear explosives for peaceful purposes. Accordingly, in order to compensate non-nuclear-weapon states for agreeing to forego the right to develop such devices, it was provided in Article V that each party to the Treaty 'undertakes to take appropriate measures to ensure that . . . potential benefits from any peaceful applications of nuclear explosions will be made available to non-nuclear-weapon States Party to the Treaty on a non-discriminatory basis'. Article V then emphasises that these benefits are to be available 'under appropriate international observation and through international procedures'.

The task of supervising such arrangements was not specifically bestowed upon the IAEA, Article V simply referring to 'an appropriate international body with adequate representation of non-nuclear-weapon states'. However, on the basis of a report by the Board of Governors, the General Conference of the Agency recommended in 1969 that it should assume this role, and this development was endorsed by the U.N. General Assembly in Resolution 2665(XXV) on 7 December 1970.

If explosive devices are to be made available by nuclear-weapon states for use on the territory of non-nuclear-weapon states, it is important in the interests of non-proliferation that adequate controls are imposed to prevent information being obtained by the latter that might assist them in developing nuclear weapons. In 1972, the Board of Governors adopted a set of 'Guidelines for the International Observation by the Agency of Nuclear Explosions for Peaceful Purposes under the Provisions of the Treaty on the Non-Proliferation of Nuclear Weapons or Analogous Provisions in Other International Agreements'. Under these guidelines, observation by the IAEA of a peaceful nuclear explosion is required where such an explosion is carried out through the Agency or pursuant to agreements made in accordance with Article V or with other similar treaty provisions. However, observation is only to be undertaken pursuant to special 'observation agreements' which are to be entered into between the Agency and the state or states concerned under terms complying with the provisions set out in the Guidelines. The underlying principle of these agreements is set out in section 8 of the Guidelines which stipulates that they are to provide assurances against violation of Articles I and II of the NPT or analogous provisions in other treaties which are intended to prevent non-nuclear-weapon states from acquiring or manufacturing nuclear explosive devices.

While the objective of these assurances is acceptable enough, it remains to be seen whether observation agreements, should any be drawn up in accordance with the guidelines, are able to guarantee that the state receiving the service will not obtain information of use to it in a weapons-making project. For the moment, it is recognised that there are grounds for concern at the possible danger of peaceful nuclear explosion projects in this respect. The position might be made clearer when an Ad Hoc Advisory Group, set up by the IAEA Board of Governors in 1975, presents its report some time during 1977.

Apart from the question of whether a state is entitled to develop and explode a nuclear explosive device for peaceful purposes, there are limitations contained in the Treaty Banning Nuclear Weapon Tests in the Atmosphere of 1963 upon how an explosion can be carried out. Although this Treaty is subject to a withdrawal clause similar to that contained in the NPT, a party to the former is not permitted, while it remains a party, to conduct peaceful—or indeed any other—nuclear explosions on the surface of its territory or under water; and it can only conduct them underground if it prevents the escape of radioactive debris beyond its jurisdiction or control.

Moves to impose more stringent safeguards

There is an increasing awareness amongst states supplying nuclear materials, facilities and information, of the link between the peaceful uses of nuclear energy and the development of a nuclear weapon-making capability. Two issues have come to be regarded as crucial: the transfer of technology (know-how), and the supply of 'sensitive' equipment (uranium enrichment facilities, or reprocessing plants for separating plutonium from irradiated fuel). Nuclear weapons proliferation is brought about as much by making available the necessary technical knowledge that inevitably passes with the supply of facilities as by misuse of the facilities themselves. This is particularly the case when it comes to the provision of enrichment facilities or a reprocessing plant.

Recognition of this basic factor has brought about a fundamental change of attitude in the United States where attempts are being made in Congress to reverse the consequences of two decades of thinking based upon the 'atoms for peace' philosophy. Furthermore the United States has been engaged in consultations since early 1975 with the governments of other suppliers of nuclear equipment with a view to placing additional restrictions upon such supplies. These discussions, carried out under the sobriquet of the London Suppliers' Club, have aimed at establishing a common policy. The original talks were held between the United States, the Soviet Union, the United Kingdom, Canada, France, the Federal Republic of Germany, and Japan, but Sweden, the German Democratic Republic, the Netherlands, Belgium, and Italy joined at a later stage.

No formal record has been published of the meetings, and most reports of what took place are largely conjectural. It does appear that the United States is seeking to incorporate arrangements, agreed to on an informal basis, in some more formal treaty. This step is being resisted by those Euratom states that are members of the Suppliers' Club. The Euratom Treaty is based upon the principle of freedom of commerce in nuclear items between Euratom states. Any formal agreement among Suppliers' Club members involving restrictions on the transfer of equipment or materials to other Euratom states would amount to a breach of the Euratom Treaty.

While, in the informal understandings among the members of the Suppliers' Club, it has been accepted that the imposition of more stringent safeguards should also cover technical knowledge made available to the recipient state, there appears to have been a fundamental difference of opinion over the supply of sensitive equipment. France and the Federal Republic of Germany have insisted upon their right to sell such facilities provided that additional safeguards are imposed.

The transaction that has perhaps caused the greatest controversy is the Co-operation Agreement of 1975 between the Federal Republic of Germany and Brazil which provides for the construction of German designed reactors, and the setting up of enrichment, fuel fabrication and reprocessing plants in Brazil, and the training of Brazilian scientists in the techniques involved in operating the various facilities. The criticisms made of this agreement, particularly by the United States, and the reactions to those criticisms by the two parties, illustrate some of the major issues being faced by the Suppliers' Club.

Brazil's attitude was based principally upon resentment at what was regarded as unwarranted interference by the United States in the internal affairs of Brazil. The Federal Republic of Germany concentrated upon the non-proliferation aspects of the situation.

It was pointed out that Brazil already had a nuclear potential based upon a program initiated in the early 1950s and developed through a number of research reactors built between 1958 and 1973. Furthermore, under a thirty year agreement signed in 1972, the United States had agreed to supply low-enriched uranium to a 600 megawatt nuclear power station being built by Westinghouse at Angra dos Reis. In the German view, therefore, Brazil was already within striking distance of a weapons making capability based upon existing technological developments and indigenous reserves of natural uranium. If Brazil was intent on acquiring nuclear weapons, it could do so whether or not Germany entered upon a huge peaceful nuclear power development program in Brazil. Moreover, whether the program would advance a weapon-making potential depended upon the extent to which it was possible to employ the facilities, equipment and technology for such a purpose, but this possibility, so the Germans argued, would be excluded by the terms of the safeguards agreement. This agreement, the draft terms of which have already been approved by the IAEA, is to apply to all items supplied under the co-operation agreement between the two states.

The safeguards are in keeping with the importance attached by all members of the Suppliers' Club to a number of more stringent requirements. The draft agreement incorporates a provision, based upon the lesson of the Indian explosion, imposing a blanket proscription on the use of any of the items listed (materials, equipment, etc.) 'for the manufacture of any nuclear weapon or to further any other military purpose or for the manufacture of any other nuclear explosive device'. It applies the prohibition to any nuclear facility 'designed, constructed or operated in one of the said states on the basis of or by the use of relevant technological information transferred from the other', and to any nuclear material 'produced, processed or used on the basis of or by the use of any relevant technological information transferred from one of the said states to the other'. The agreement is potentially of indefinite duration. Even if it ceases to be in force because there is no nuclear material left that is subject to safeguards, the agreement will immediately revive if 'a nuclear facility or specified equipment is designed, constructed or operated on the basis of or by the use of relevant technological information transferred from the other'. There are restrictions upon retransfer without the consent of the supplier state; these restrictions are amplified in the co-operation agreement.

Despite these welcome features incorporated in the safeguards arrangements, American fears have not been assuaged. This apprehension is based upon the view, which is increasingly being adopted by an influential body of opinion in the United States, that governments are not to be trusted with sensitive technology, however stringent the safeguards requirements might be. The impracticability of determining whether a transfer of technology, as opposed to a new development, had actually taken place supports this view. Every diplomatic effort is being made to prevent sales of enrichment and reprocessing facilities, and support is being given to the idea that such facilities should be established on a regional basis under international controls.

No discussion of the move towards restricting the freedom of nuclear commerce would be complete without some reference to the more fundamental issues of how far this development is compatible with the NPT. In one obvious respect, this change in attitude is in contrast to the extension of the peaceful uses of nuclear energy envisaged by Article IV of the Treaty. However, as long as reactors, the basic essentials of a peaceful energy program, are provided to developing countries wishing to acquire such facilities, a policy that merely seeks to retain control over sensitive technologies and to impose more stringent safeguards may be consistent with the Article.

The underlying conflict is of a different character. It is that states have become parties to the NPT on the basis of certain safeguards requirements set out in Article III and subsequently elaborated upon in agreements between themselves and the IAEA. The chief proponents of the NPT, the United States, the Soviet Union, and the United Kingdom, are now in effect saying in their capacity as leading figures in the Suppliers' Club negotiations that these safeguards are inadequate. Not only must Article IV be more restrictively interpreted, but Article III must be replaced by a more complete set of safeguards requirements.

Mention has already been made of the fact that, although the NPT requires the application of safeguards to all source or special fissionable material, the actual provisions contained in agreements under the NPT do not in fact require the safeguarding of source material. Safeguards therefore do not hinder a country wishing to develop nuclear weapons from obtaining the source material from which nuclear explosive material is derived. The extension of safeguards to cover source material is clearly desirable, although we recognise that this would necessitate a very large increase in the safeguards systems, in this cost, and in the inspectorate required.

Conclusions

The main limitations and weaknesses of the present safeguards arrangements can be summarised as follows: the failure of many states to become parties to the NPT; the inability of safeguards to prevent the transfer of nuclear technology from nuclear power production to the acquisition of nuclear weapons competence; the fact that many nuclear facilities are covered by no safeguards; the existence of a number of loopholes in safeguards agreements regarding their application to peaceful nuclear explosions, to materials intended for non-explosive military uses, and to the retransfer of materials to a third state; the absence, in practice, of safeguards for source materials; the practical problems of maintaining effective checks on nuclear inventories; the ease with which states can withdraw from the NPT and from most non-NPT safeguards agreements; deficiencies in accounting and warning procedures; and the absence of reliable sanctions to deter diversion of safeguarded material.

The Commission recognises that these defects, taken together, are so serious that existing safeguards may provide only an illusion of protection. However we do not conclude that they render valueless the concept of international safeguards. We believe it is both essential and possible to make safeguards arrangements more effective.

Many countries, including some states with the capacity to develop nuclear weapons quickly, have not become parties to the NPT. The Commission concludes that all possible steps should be taken by nuclear-weapon states to induce a change in attitude of non-party states towards the Treaty. These should include a greatly increased effort to implement Article VI of the NPT, relating to disarmament.

A major concern about the ability of the NPT to prevent further proliferation of nuclear weapons among countries arises from Article IV of the Treaty, which upholds the right of all parties to participate in the fullest possible exchange of nuclear equipment, materials and scientific and technological information. This exchange inevitably also spreads the means to develop a nuclear weapons capability.

Any nuclear resources transferred by one state to another should be subject to international safeguards. Imposition of IAEA safeguards on a particular facility provides a degree of assurance against diversion. However, other facilities which are free from safeguards may already exist or may be developed in the same country. Before delivering nuclear items, the transferring state should see that full safeguards are imposed on all nuclear resources existing in, or to be acquired by, the recipient country. Safeguards should be extended in practice to cover source material (including yellowcake).

Agreements on supply of uranium entered into by Australia should incorporate an undertaking that nuclear material will not be used for military purposes. This action would overcome a weakness in the NPT whereby a recipient state party to the Treaty may withdraw material from safeguards by claiming to use it for non-explosive military purposes.

Australian uranium should not be retransferred by a recipient state to a third state under conditions less stringent than those imposed by Australia on the first recipient state. Means are available for imposing safeguards on retransfers. The Commission believes that a clause in the treaty with the original recipient state prohibiting retransfer without Australia's approval would be the most satisfactory solution.

Doubts about whether the development of nuclear explosive devices for peaceful purposes is prohibited by IAEA safeguards agreements negotiated before the Indian nuclear explosion in 1974 constitute a weakness in those agreements. There is essentially no difference between peaceful nuclear explosive devices and atomic bombs. The extent of this problem has been reduced by the inclusion in post-1974 IAEA safeguards agreements of a specific proscription on the use of materials for any nuclear explosive device. The IAEA has produced guidelines for agreements to cover the provision of peaceful nuclear explosive devices to non-nuclear-weapon states under Article V of the NPT. These should enable any benefits offered by such devices to be made available to countries prohibited from developing them. It seems unlikely that a proliferation threat will arise from the provision of peaceful nuclear explosive devices under IAEA guidelines in the foreseeable future. However, unilateral development of such devices using unsafeguarded material, as was done by India, is a continuing risk.

The Commission acknowledges disquiet expressed about the ease with which a state can withdraw from the NPT. There appears to be no solution to this problem. States are not prepared to commit themselves permanently to the non-development of nuclear weapons in case their security interests might require a change of policy. Fuller implementation of Article VI of the NPT might make states more willing to adhere to the Treaty, but it would be unlikely to affect their attitude to the right of withdrawal.

The Commission recognises the possibility that a state supplied by Australia with nuclear materials might withdraw from the NPT or from IAEA safeguards or from both. We conclude that nuclear materials should be supplied to a state only on the basis that its entire nuclear industry is subject to back-up safeguards that cannot be terminated by unilateral withdrawal.

Accounting procedures are the principal means of determining whether nuclear material has been diverted from safeguarded facilities. Existing accounting procedures are not sufficiently accurate to provide adequate assurance that diversion of fissile material would be detected. The technical and administrative procedures for detection and confirmation are so involved that they would not provide a warning in time for the international community to do anything about a violation.

There is widespread disinclination among states to accept regular inspections, on the grounds that such inspections are incompatible with national sovereignty. The Commission notes difficulties in maintaining suitable personnel to carry out inspection routines and also the problems of financing such surveillance. Problems with inspection will become much greater if projected expansions in nuclear power capacity take place.

The limitations of the sanctions against diversion specified in the IAEA Statute reflect difficulties with enforcing obligations that underlie the entire international legal order. While a solution to this general problem may not be attainable, the force of sanctions related to materials supplied by Australia would be increased if supply were made through countries such as the U.S.A. on which recipient countries may be highly dependent. For example, arrangements might be made for Australian uranium to be enriched in the U.S.A. before delivery to other purchasers.

Plate 10. Spent fuel elements can be seen in this cooling pond at a reprocessing plant at Morris, Illinois, U.S.A. Reprocessing was scheduled to begin at the plant in 1974. However, because of technical difficulties requiring extensive modifications to the plant, reprocessing is not now expected to begin for several more years. (Photo by courtesy of the U.S. ERDA.)

Plate 11. Spent Magnox fuel elements being transported in a heavy, cooled container away from Oldbury nuclear power station in the U.K. (Photo by courtesy of the Central Electricity Generating Board.)

Plate 12. The Windscale nuclear complex in the U.K. Installations include the Calder Hall Magnox power reactors (on the right; their cooling towers are shown releasing steam to the atmosphere), a prototype advanced gas-cooled reactor (on the left), fuel fabrication and reprocessing plants (in the central group of buildings), and storage tanks for liquid high level waste. A pipeline carrying low level liquid waste from the reprocessing plant to the sea can be seen on the left. (Photo by courtesy of British Nuclear Fuels Ltd.)

14 NUCLEAR THEFT AND SABOTAGE

Concern was expressed during the Inquiry about the theft and illicit use of nuclear materials and the sabotage of nuclear facilities. The discussion centred around action by terrorists, but covered also the potential involvement of criminal groups and malicious individuals.

Activities of terrorist organisations include attempts to draw attention to a political cause, to gain an offensive capability, to attack politically sensitive targets, to precipitate civilian disruption, and to create blackmail situations. Nuclear power systems are obvious foci for such motives, being regarded variously by terrorist organisations as a symbol of national prestige, as a source of aggressive potential, and as being contrary to their ideological beliefs.

The potential action which caused most concern at the Inquiry was theft of nuclear material, especially plutonium, which could be used for weapons or as a radiological poison. Another threat was sabotage of nuclear plants causing destruction, a radiation hazard to surrounding populations and costly disruption of power systems. This category included terrorist occupancy of a nuclear power station—an event which has already occurred in Argentina.

Although existing terrorist organisations operate largely outside Australia, there are global ramifications. The Commission was informed during the sole private session of the Inquiry about international terrorist organisations, their capacity for terrorist acts and their modus operandi, and was informed that steps are taken to keep under surveillance the activities of such organisations as they affect Australia.

Organised crime has demonstrated a high degree of organisational and technical competence, especially in the U.S.A., for theft involving complex technological systems. Evidence was given to support the view that organised criminal intervention was a likelihood should the financial rewards be sufficiently great. Some envisaged a nuclear 'black market' where contracts were placed by, say, terrorist groups with criminal organisations for theft of nuclear materials or plant. A Commissioner of the U.S. Atomic Energy Commission has stated that the development of a blackmarket in plutonium is likely.

Finally there was the prospect of action by an individual with malicious intent, motivated by the prospect of financial gain, ideological beliefs, or anti-nuclear convictions or suffering mental derangement. Such a person would have scope for small-scale diversion of harmful materials and, if technically knowledgeable, a considerable potential for sabotage. If coerced or otherwise incorporated into a terrorist group, an expert person, especially an employee of a nuclear installation, would greatly increase the risk of serious incidents.

The Commission was given evidence about means to prevent theft and sabotage. Naturally, the matter could not be examined in full because of secrecy regulations on surveillance and protection. Numerous breaches of security in nuclear installations have been recorded, and there was evidence that attempts were being made in some countries to render nuclear operations more secure. Nevertheless, the Commission does not feel confident that nuclear facilities would currently withstand determined assaults by terrorist organisations. It seems doubtful whether, as the number of facilities increases, it will be possible to provide sufficient defences to render every installation safe against attack by even small numbers of well-armed, trained men.

Theft of nuclear materials

There are many materials in nuclear fuel cycles that could possibly interest terrorists. Significant materials are plutonium, uranium highly enriched in uranium-235 and uranium-233. All these can be exploited for making weapons; plutonium has additional potential as a radiological poison.

Plutonium is contained in all spent fuel elements which contained uranium-238. The extreme radioactivity of the spent elements seems sufficient to protect them against seizure by terrorists until reprocessing is complete—when the intensely radioactive fission products, and the other actinides, have been separated from the uranium and plutonium.

Large quantities of plutonium already separated in reprocessing plants are currently stored in the form of a liquid plutonium nitrate solution, waiting to be recycled through reactors. An OECD-NEA/IAEA estimate, based on the assumption that reprocessing plants will start operating and plutonium recycling will occur, puts annual plutonium extraction from world nuclear power programs in 1980 at some 18 tonnes, rising to 109 tonnes by 1990.

At present, some 7 tonnes of plutonium from civilian power reactors are stored in the U.S.A. If reprocessing of oxide fuels resumes, this quantity will increase annually, even after recycling comes into operation. Other countries which plan to recycle plutonium through reactors holds stocks of plutonium. There are unspecified quantities stored in nuclear weapons establishments throughout the world and several hundred tonnes of plutonium are contained in war-heads of atomic weapons under the control of armed forces. Doubts about the wisdom of recycling plutonium are widespread, because of both the dangers inherent in the routine circulation of the material and the rapid increase in recent estimates of the cost of reprocessing. The U.S. Government does not intend to make a decision before 1977 on whether or not to introduce plutonium recycling.

Plutonium nitrate solution can be readily converted to plutonium oxide. Plutonium-239 in metal or oxide form is highly efficient explosive material; only about 6 kilograms of the metal are needed to produce an explosive chain reaction, and only about 9 kilograms of oxide. However, the plutonium that comes out of a reactor contains other isotopes besides plutonium-239, particularly plutonium-240, which to some extent reduce the usefulness of the plutonium for weapons. The proportion of isotopes other than plutonium-239 depends on the length of time the fuel has been in the reactor, and other factors (see Chapter 12).

The evidence suggests that the most serious likelihood of diversion will arise if reprocessing and recycling of plutonium are established on a large scale. This will be essential if fast breeder reactors come into regular service. Plutonium nitrate will then be commercially converted to the oxide and used on a large scale as reactor fuel. In fuel form, plutonium oxide will usually be mixed with other oxides of fissile or fertile nuclides. Although it is possible to separate plutonium from mixed oxides, it is easier technically to make plutonium oxide from plutonium nitrate solution.

Enrichment is the first step in a fuel cycle based on uranium-235 to produce material of possible interest to terrorists. However, only highly enriched uranium can be used for bomb-making; the slightly enriched fuel used in most power reactors is as incapable as natural uranium of being used for that purpose.

Highly enriched uranium is used in the one high temperature gas-cooled reactor (HTGR) operating commercially, and may be used in fast breeder reactors (FBRs). But at present very little is produced for non-military purposes. A possible target in the production of fuel using highly enriched uranium would be the oxide (or carbide) produced from uranium hexafluoride prior to incorporation into fuel elements. Uranium hexafluoride could also be stolen from enrichment plants or *en route* to fuel fabrication units. However, hexafluoride could not be used directly for making bombs, but would have to be converted to uranium metal (or possibly oxide) in a clandestine laboratory. Evidence indicates that breaking down HTGR fuel elements to recover uranium suitable for constructing weapons would be exceedingly difficult.

Uranium-233 is produced in reactors with thorium in their fuel—at present only the HTGR. It could be separated during reprocessing of spent fuel from these reactors, but this process has not yet been commenced. Uranium-233 has explosive potential comparable to plutonium, but is a less dangerous material with which to work. It represents a potential threat for the future depending on the extent to which thorium is included in reactor fuels and on the introduction of reprocessing of these fuels.

The construction, use or threat of a terrorist bomb

A number of witnesses treated the possibility of terrorists making a nuclear explosive device as remote, and suggested that it would be easier for them to acquire an existing military weapon. The weight of evidence available to the Commission suggests that a terrorist team could, if conditions favoured them, construct a very destructive device. The bomb-makers would have to be able to work undisturbed for weeks or possibly months, although most of the work could be done before the explosive material was acquired. They would need to call on the assistance of at least one person with the necessary laboratory skills and sufficiently well versed in the published literature dealing with nuclear explosives.

As the amount of highly-enriched uranium in use and circulation is limited, and uranium-233 is not yet being produced, the most vulnerable material for the immediate future would appear to be plutonium. As most of this is now stored in nitrate form, the laboratory would have to be able to convert it at least to plutonium oxide. Illicit bomb-makers would have to be on guard to avoid the accidental assembly of a critical mass. To avoid personal contact with toxic airborne particles, the plutonium would have to be handled most carefully in airtight enclosures. However, it was pointed out that self-preservation was not a notable characteristic of terrorist behaviour.

The evidence before the Commission suggests that a terrorist group could use reactor grade plutonium to make a bomb with good prospects of giving a yield of several hundred tonnes of TNT. Although there would be considerable uncertainty about the yield before the bomb was actually detonated, this is likely to be of little concern to a terrorist. An explosive yield of a few hundred tonnes of TNT might be sufficient to destroy a very

24647/76—11

large skyscraper, with severe loss of life. The ionising radiation released and the subsequent fall-out would also kill and injure many people.

It was also suggested that a bomb of this kind might possibly explode during manufacture, doing widespread damage, but also killing the makers. However, this risk does not detract from the real possibility of manufacture of an explosive device by terrorists to whom the risk of death might be an insignificant consideration compared with the spectacular manifestations of success, should that be achieved. The length of time and the sophistication of laboratory facilities needed for the operation would depend largely on the nuclear materials available.

The mere possession of nuclear materials coupled with the threat or pretence of bomb-construction may suffice to produce the effect desired by terrorists. It was suggested that terrorists might threaten to explode a nuclear device if their demands on a government were not met. The Commission was told that, at the current ransom rate for kidnappings, a terrorist group threatening a nuclear explosion might escalate its demands to many millions of dollars.

A terrorist organisation might stipulate some political objective as the 'ransom' to be paid. It can be asked, for example, what the reaction of the U.S. or British governments would be if a terrorist group threatened to explode a nuclear device in New York or London if, respectively, Israel did not withdraw entirely from the occupied territories so that a Palestinian state could be created, or British troops were not withdrawn from Northern Ireland to allow integration of the province with the Irish Republic.

In any situation in which a threat was made to set off a nuclear device, the security forces and the government concerned would face an obvious dilemma; how seriously should they take the threat? On the one hand, there would be the incalculable consequences if the threat materialised. On the other, the authorities would be aware of the need to hold out against the possibility of bogus threats. It would be very easy for any individual or disaffected group to advance demands on the pretention of a nuclear threat.

Some witnesses suggested that, to an extent, the authorities could rely on the supposition that, as the terrorists presumably would wish to achieve their objective, they would be prepared to offer evidence of the existence of their explosive device. Examples given of the form this evidence might take include a sample of the plutonium used, or details of the plutonium and how it was obtained, and of the bomb and how it was made.

Production by the terrorists of some plutonium would be much more convincing evidence than the production of technical details, although it is conceivable that a group which had produced a bomb might not be able to provide excess plutonium on demand for logistic reasons or, alternatively, that a group which had not produced a bomb could provide some plutonium if any had in fact been diverted. Design features produced would almost certainly be copied from available literature whether or not the bomb actually existed. It is concluded that the authorities probably would be in no better position to judge after the information or material was given than before whether the threat was real. Similarly, details supplied about the plutonium and the method used to obtain it would probably not enable the authorities to determine in the time required whether a clandestine diversion had in fact occurred. If the alleged source of the plutonium was in another country, there would be even less chance of making a reliable assessment of the likelihood that the threat was genuine.

Use of plutonium as a radiological poison

The prospect was raised in evidence that terrorists may choose to disperse powdered plutonium into the atmosphere. Here again the threat of such action might suffice to produce the desired effect, especially as the harmful consequences of plutonium inhalation are not immediately fatal or disabling. Unless quantities sufficiently large to cause direct lung damage are inhaled, the consequences become apparent as cancer only after a period of years.

Plutonium oxide exists as a powder. It may be derived by adding sodium oxalate, a widely used chemical, to plutonium nitrate solution stolen from a storage unit. Plutonium oxalate is filtered off and heated to form plutonium oxide powder. It was stated in evidence that finely divided plutonium metal will spontaneously burst into flame in air to form ideally dispersible particles of plutonium oxide.

Some witnesses saw problems with effective dispersal of plutonium particles in confined spaces. If projected by explosives, plutonium powder would be subject to prevailing meteorological conditions which could render uncertain any planned dispersal. Also, a cloud of plutonium particles could be a hazard to those who released it, although this may not concern terrorists. Any actual release, whatever its radiological effects on the population, would provide authorities with the additional and difficult task of evacuating and de-contaminating an exposed area.

Other witnesses believed that terrorists would prefer some other more readily dispersible agent to plutonium, such as chemical 'nerve gases'. Be this as it may, the atmospheric dispersal of powdered plutonium remains a potential threat, particularly in view of the possibility that in the future it may be handled routinely in large quantities.

Prevention of nuclear theft

The Commission was presented with considerable evidence on the subject of preventing illicit diversion of nuclear materials. Discussion dealt mainly with the theft of plutonium. Attention was concentrated on areas vulnerable to illegal access—the chemical processing plants of the nuclear fuel cycle and material in transit, accessible in power stations or accessible in storage units. The Commission was informed of numerous incidents where nuclear materials had been stolen, were lost or simply could not be accounted for. There were cases where concerned individuals deliberately tested security measures, as when a German parliamentarian carrying a 60 cm bazooka entered the world's largest nuclear power station, unhindered. It was apparent that there were notable deficiencies in arrangements for protecting nuclear facilities. But it was equally apparent that, in recent years, authorities have been making greater attempts to remedy existing weaknesses.

Some witnesses claimed in evidence that the safeguards against diversion based upon accounting are inadequate (see Chapter 13). Regulations with provisions aimed at preventing break-ins into nuclear facilities have been established for some years by the International Atomic Energy Agency (IAEA) and individual countries with nuclear power facilities. A recent report by the U.S. Government Accounting Office revealed very substantial deficiencies in compliance with such regulations in the United States, and significant corrective measures have been taken.

There has been special concern in the U.S.A. about transportation arrangements, due in part to revelations of casual and unguarded trucking of plutonium nitrate and also to concern about the hijacking skills, and the degree of control of the trucking industry, acquired by organised criminals in that country. As a result, regulations have been recently introduced for guarding nuclear materials in transit.

The IAEA has set out recommendations on requirements for the physical protection of nuclear materials in transit internationally. These suffer from the disadvantage of not being enforceable. However, there is a strong current of international opinion in favour of increased physical security for nuclear material.

Whatever the degree of physical surveillance established, however, the evidence suggests that, given the element of surprise, an attacking force consisting of as few as three well-armed, trained men would be difficult to thwart. The long history of aircraft and vehicle hijackings illustrates the problem. The weakest point in the transportation process is likely to be trans-shipment from one form of transport to another. This is one reason for the proposal that 'nuclear reserves' be created to contain in the one area all processing units of the fuel cycle as well as power stations. Attractive though this may be for reasons of security, there are disadvantages in terms of military strategy and distribution of a country's energy resources.

From a terrorist's standpoint, clandestine diversion clearly would be a safer method of obtaining nuclear material than a break-in or hijacking. Witnesses expressed concern that measures needed to minimise the chances of diversion might constitute an infringement of civil rights. These, it was suggested, would include investigations of the personal attitudes of employees and of applicants for jobs. Some witnesses contended that, should the use and transportation of plutonium become a regular feature of the nuclear industry, the degree of investigation and subsequent surveillance which would be necessary would be unacceptable in a democratic society.

To an extent, this issue is a matter of individual choice. Acceptance of employment in the nuclear industry certainly involves acceptance of a degree of supervision and investigation. Employment in the armed services was pointed to as an analogy; a person entering the forces could not opt out of the limitations placed upon his way of life by the exigencies of the service. If Australian uranium is only mined and milled for export, the amount of additional security control manifesting itself in the Australian community is likely to be miniscule as no material of value to terrorists would be produced in this country. But if later stages in the fuel cycle were developed here—particularly enrichment facilities, reactors and a reprocessing operation—strict security control could be expected covering plant, storage areas, and the transport of nuclear materials.

Sabotage of nuclear facilities

The Commission heard evidence on the likelihood that terrorists or malicious individuals would attempt or threaten to blow up nuclear power stations or other facilities in the fuel cycle. It was agreed that these possibilities had to be seriously considered, and attention centred mainly on attacks against nuclear reactors in power stations. Some witnesses believed that the physical containment of reactors would withstand attacks with conventional explosives or the deliberate crashing of aircraft. Other evidence indicated that varying degrees of vulnerability were associated with different reactor designs.

If the aggressors included a nuclear specialist with knowledge of safety systems, including fault-tree analyses, they might design an attack which would by-pass the sequential protective mechanisms engineered into a system aimed at preventing the spread of a potentially dangerous malfunction under normal conditions of operation. Such expert assistance would be necessary if the intention was to achieve the maximal destructive effect—to initiate a loss-of-coolant accident and to breach the reactor containment of steel or concrete. This could lead to reactor core meltdown and escape of radioactive materials through broken coolant pipes and breached containment shields (see Chapter 10).

Detailed consequences of an attack remain conjectural, due to lack of any comparable incident. Reactor design is important. For example, older gas-cooled units may have relatively more accessible coolant pipes and heat exchangers separated from the main reactor building. An explosion in the heat exchanger of a gas cooled reactor could conceivably force steam and water into the core and possibly induce further local explosions. It needs to be reiterated that no form of attack on a thermal nuclear reactor can result in an atomic explosion, but the possibility remains that, through skilful sabotage, the reactor site could be damaged and the surrounding community could be seriously harmed by the release of great quantities of radiation. Some witnesses suggested that an inefficient but destructive nuclear explosion could occur in an FBR if actions were taken to cause the core to melt and the fuel to agglomerate. However, methods of preventing such agglomeration have been devised, and these should minimise, and perhaps eliminate, this possibility.

There are opportunities for causing great damage by sabotage of nuclear installations other than reactors. Cooling ponds for spent fuel at reactor sites or reprocessing plants contain highly radioactive material which, if dispersed by a large explosion, would cause serious local hazards. Such risks also apply to other parts of reprocessing plants and to fuel fabrication facilities handling plutonium. Sabotage of spent-fuel transportation casks in transit would be difficult because of the rugged construction of the casks. It would also be highly dangerous to the saboteurs (see Chapter 10). But if it occurred, it could cause a local problem over the area of dispersal. Also, there are theoretical possibilities of serious radiation hazards if high-level waste storage tanks are breached.

As with measures for preventing theft of nuclear materials, there have been recent moves to improve the protection of nuclear facilities from sabotage. Nevertheless the Commission formed the opinion that a determined and well-organised attack, especially if assisted by a nuclear expert, could be mounted with a realistic chance of causing very serious damage.

General measures for controlling terrorist groups

Terrorist organisations operate in a variety of ways. Their position is obviously enhanced if they can operate from a base within the local community, as with the Mafia in the U.S.A. or the Irish Republican Army in parts of Ireland. However, even where a terrorist group can rely on no connection with the local inhabitants, the relative freedom of travel between one country and another and the availability of false travel documents make the task of adequate surveillance very difficult.

Experience suggests that terrorist groups are much more likely to be caught or neutralised when they repeat a pattern of action than the first time they act. Unfortunately, in the field of nuclear terrorism, an isolated incident could constitute a major catastrophe. It was suggested that a few nuclear attacks by terrorists, or attacks on nuclear facilities—particularly if one were to involve a disastrous escape of radioactive materials—could lead to a total reversal in energy policy among Western countries and involve the curtailment of nuclear power programs.

Most discussion of the terrorist threat has dealt with the activiites of organisations with relatively well-defined objectives and exercising a degree of control over their members. However, it is possible to envisage various other situations. A group might be encouraged to act as a surrogate army on behalf of a state wishing to produce the maximum destruction in the territory of an 'enemy' state without going to war itself. Or an anarchist or extreme element in a terrorist organisation might seek the ultimate in destruction in a state opposed to the organisation's aims.

The problem is increased by the fact that a terrorist group is freed from a major restraint that operates between nations—the fear of nuclear retaliation. As the group has no territory of its own, it cannot be singled out for such retribution by the injured state or its allies. This would be the case even if the group was acting clandestinely on behalf of a state.

Within any country, the possibility exists that dissident groups or disgruntled individuals might turn their antagonisms to nuclear sabotage. Awareness of this problem has led to the imposition of greater physical security requirements in the United States and elsewhere. It was pointed out that there is a psychological side to the issue. Apart from targets selected for purely personal motives, a protest against society of this sort is more likely to involve an object that is symbolic or newsworthy, such as a nuclear plant.

Attempts made in the United Nations to produce a general convention binding nations to take action to prosecute, or hand over for prosecution, those responsible for terrorist acts have made little progress. An influential argument against such a convention has been that the principal form of 'terrorism' is that employed by states to persecute minorities, and that indiscriminate acts of violence by members of such minorities struggling against 'colonial' or alien domination are in some way justified.

It was suggested to the Commission that it might prove possible to draw up a more limited code of legal rules at the international level. In the way that there are conventions dealing with hijacking and other acts directed against civil aircraft, it could prove possible to proscribe all acts of terrorism involving nuclear material or facilities.

Conclusions

'Weapons grade' plutonium can be produced in most power reactors by operating them in a manner which is not compatible with the most efficient generation of electricity. However, the evidence points strongly to the conclusion that very destructive nuclear explosive devices could also be made from 'reactor grade' plutonium produced in power reactors operated normally. The physical data needed to make such devices, with uncertain explosive yield but probably in the range of hundreds to thousands of tonnes equivalent of TNT, are available in the open literature. Construction of a nuclear explosive device is not of such complexity as to be beyond the apparent resources of existing terrorist organisations. Production of a device to disperse plutonium oxide in the atmosphere would be comparatively simple. The evidence indicates that undetected theft of small quantities of plutonium from reprocessing plants would probably be feasible, especially for trained organisations in countries with large nuclear industries.

Plutonium is the material of most concern at the present stage of development of the nuclear power industry, and the risk of its theft will increase markedly if plutonium recycling is introduced on a commercial scale. If fast breeder reactors are to fill the role envisaged for them, recycling will be necessary. Nuclear bombs can also be made from uranium-235 and uranium-233. In the future these materials may be produced in relatively pure form for fuelling reactors. If this occurs, the risks of theft will be similar to those described for plutonium.

An attempt by even a small, well trained and armed group to take over a nuclear installation could have a good chance of success. Subsequent threats to destroy the installation and release large quantities of radiation would have to be taken very seriously indeed.

There is a very real risk that the opportunity and the motive for nuclear blackmail will develop with time. Some common characteristics of terrorist groups suggest that they might attempt to make and explode atomic bombs or make other terroristic uses of nuclear materials or facilities. These include lack of concern for their own safety and for the suffering caused by their attacks. Since they have no territory of their own, fear of nuclear retaliation is not an inhibiting factor. Major difficulties could arise in attempting to determine the reality of a threat by a group to explode an atomic bomb, to spread radiation from a reactor, or to disperse plutonium. Either acceding to or refusing the demands of such a group could have very adverse consequences for society.

Measures designed to prevent theft of nuclear materials and attacks on nuclear installations have been tightened in recent years. Welcome as those measures are, the evidence indicates that the risks are presently real and will tend to increase with the further spread of nuclear technology.

Plate 13. Part of an abandoned salt mine in Kansas, U.S.A. Places like this are among possible sites for the permanent disposal of solidified high level waste. This mine has been used in tests on the feasibility of using such salt deposits for this purpose. (Photo by courtesy of the U.S. ERDA.)

15 FUTURE ACTION

Nuclear energy and alternatives

In this Chapter we first examine the need for the countries of the world to rely upon nuclear energy. Having the view, as we do, that all reasonably practical steps should be taken to limit reliance on nuclear energy, we conclude by making some proposals to that end.

The action by the Organisation of Petroleum Exporting Countries (OPEC) in late 1973 in quadrupling the price of oil caused a sudden and massive deterioration in the trade balances of countries heavily dependent on imported oil. This, and the earlier embargo imposed on the supplies of oil to certain countries by the Arab members of OPEC, caused intensive re-examination of the bases on which the importing countries had been proceeding in obtaining and using energy. The vulnerability of modern, technically advanced economies to changes in the price or availability of energy resources became almost universally recognised. Government policies in those countries were and are now being developed which in general seek to reduce dependence on imported oil.

The studies which sprang out of the oil crisis confirmed the finite extent of the energy resources on which industrialised economies largely depend, and the comparatively short times in which some resources, particularly petroleum, were likely to be exhausted if energy use throughout the world continued to increase exponentially at the rates prevailing before 1974. Examination was made of the need, in economic and social terms, for this continued increase in energy consumption.

It is argued by some proponents of nuclear power that continued growth in economic activity and consequent energy use, even if at somewhat lower rates of increase than previously experienced, is essential to the well-being of nations and that the rapid development and deployment of nuclear energy is necessary if major global economic and social crises are to be avoided in the next two decades. Opponents of nuclear power argue that continued growth in the rate of energy use, particularly in countries which are already industrially developed, will lead to such adverse environmental and social effects that it should be deliberately restrained. They maintain that the deployment of nuclear energy in particular is undesirable not only because of the dangers inherent in the technology, but because it will encourage the continued profligate use of energy by the developed nations and delay the inevitable transference of reliance to renewable energy sources, such as solar energy, thereby increasing the difficulties involved in such a transfer and exacerbating the differences in income between the developed and developing countries.

There can be no doubt that energy crises are possible in such countries as Japan and Italy, which lack adequate domestic energy resources and are dependent on imported oil for a large part of their energy production. The risk of a crisis is real in that such countries are particularly vulnerable to the actions of governments of petroleum exporting countries and therefore cannot look ahead with confidence to the uninterrupted supply of a vital ingredient in their economic welfare. Other developed countries, such as the

U.S.A., and the Federal Republic of Germany, while possessing large total energy resources, have become dependent to a high degree on imported oil and are correspondingly vulnerable.

Since 1973, governments of the developed countries have been working on energy policies designed to reduce their dependence on oil imports partly by increasing the use of alternative energy sources, and partly by developing energy conservation programs. Nuclear energy appears as an economic source of base load electricity in many parts of these countries. It is to them that future exports of uranium from Australia would most probably be directed.

The increases in oil prices and the reduction in oil resources in the next two or three decades also constitute a threat to the economically less developed countries with few domestic natural resources. Most of these countries are in the tropics and the poorest of them suffer from the additional and interrelated problems of overpopulation, excessive birth rates, and periodic food shortages. Some, but not all of them, are also lacking in technical expertise. For such countries liquid fuels are of almost irreplaceable value. Most of these countries do not have extensive electricity grids and unless, and until, massive investment in their construction is made, electricity can benefit only limited, mainly urban sectors. Because refined oil products can be used as a direct source of energy for work or heat, are easily transportable and can be utilised without major capital investment in equipment, oil is the most suitable energy source for them in making the transition from reliance on human, animal and plant energy to the development of modern, energy intensive industrialised economies to which they aspire. These countries have also become dependent in varying degree on the fertiliser derived from oil for increasing their food production. It is unlikely that they would want substantial quantities of Australian uranium in the next two decades or so because of the limited suitability of nuclear power for meeting their electricity needs.

The OPEC countries are not without problems. The major producers are in general poor in known natural resources other than oil, and are dependent to a high degree on imported food, industrial products and technology. Some of them will experience an almost total loss of export revenue when their oil reserves are exhausted, unless they develop alternative sources of export income. It is not surprising that their governments have acted to reduce the rate of depletion of their major asset by increasing its price, while they endeavour to develop alternative means of generating adequate income and living standards for their people. OPEC countries could, in the coming decades, be potential purchasers of Australian uranium, although nuclear power may not be economically competitive in these countries until their oil resources are nearing depletion.

In Australia's case, the indications are that our dependence on imported oil will start to increase rapidly before long. Mr Justice Collins has said in his *Fourth Report of the Royal Commission on Petroleum:* 'by 1984-85 indigenous crude will supply between 32 per cent to 45 per cent of the necessary crude . . . and by 1990 indigenous supply as presently estimated will be down to about 15 per cent'. The statement is based on the assumptions that annual petroleum consumption will continue to increase and that indigenous production will diminish after about 1978.

In the past, the advances made by science and technology and the incentives provided by market mechanisms and government action have combined to stimulate development of replacement resources for those which have become depleted. There does not seem much doubt that science and technology will eventually make it possible to utilise other sources of energy to replace oil. However, it is now widely accepted that the ratio of the rate of use of oil to total world oil resources may become so high that it may not be possible to develop replacement energy sources in some areas of the world in a time short enough to avoid major shortages of fuel for transport and other industries currently dependent on petroleum. Mr Justice Collins has said: 'By the end of this century world supply will be failing and ever higher prices will prevail'. If replacements are not found, either in alternative forms of energy or in methods of transport, there will be serious effects on economic and social welfare.

The oil crisis illustrates the type of problem which modern industrial societies tend to create. They are essentially problems of distribution, time and environmental quality. It is accepted that the time required for the development of a major technology from the prototype state to where it is contributing on a large scale to society's demands is in the order of two to three decades. Nuclear energy is a good example. The first commercial power reactor to feed electricity to a grid commenced operating in the U.K. in 1956. In 1974, eighteen years later, the country most dependent on nuclear energy was Switzerland, 17 per cent of whose electricity was provided by nuclear power. The U.K. was the second most dependent, relying on nuclear power for 12 per cent of its electricity. Nuclear power contributed only 3.7 per cent of the world's electricity and 1.1 per cent of the total energy used in that year. Problems arise because modern industrial societies consume at such a high rate that a critical resource may become virtually exhausted before an environmentally acceptable technology based on an alternative resource can be deployed on a sufficient scale to avoid a major economic dislocation. In order to avoid such a dislocation, society may be forced to resort to a technology which has such serious environmental and social disadvantages that a society not in the grip of a crisis would reject it.

The evidence indicates that the increased cost of petroleum, its potential exhaustion, and the possibility of sudden cessation of supplies due to political action, are at present leading some developed and developing nations into a situation where they are turning, sometimes with grave doubts, to reliance initially on nuclear power generated by thermal reactors and ultimately to nuclear power generated by fast breeder reactors. We view this trend with great anxiety. The dangers inherent in the spread of nuclear technology, particularly those associated with the routine production of large quantities of plutonium, are such that we believe it should be consciously avoided unless there is no reasonable alternative.

The question then arises, is there a reasonable alternative? The situation varies in relation to different countries and according to the considerations which are taken into account. To many developed countries, nuclear power appears as a desirable diversification in their energy programs. It permits countries dependent on imported fuels to diversify both in type of fuel and in supplier. It allows countries with their own energy resources to avoid excessive dependence on a single industry and accordingly increases security of supply. In addition, in parts of some developed countries nuclear energy appears to be the most economic source of base load electricity. In the ordinary course of events, these factors, assisted by the commitment of the established nuclear industry, could be expected to lead to nuclear energy becoming a major source of electricity in developed countries. Opponents of nuclear energy suggest, however, that in most developed countries, if not all, a 'bridge' to the general use of renewable energy sources could be provided by relying on fossil fuels and by conservation of energy. In Australia's case, no electricity generation authority currently has plans to utilise nuclear energy. Recent studies have shown that nuclear power stations are unlikely to be competitive with coal-fired stations in Australia in the foreseeable future.

We have discussed energy use in Chapter 5. It is apparent that there is a great deal of scope for reducing the rate of energy consumption required for a given level of economic activity, particularly in such countries as the U.S.A. and Canada which have historically had access to cheap energy and have therefore not been economically motivated to conservation. It is also clear that world energy resources are not in danger of imminent exhaustion. Total world coal resources are so large that they will not be approaching depletion for many decades, even if the rate of energy use continues to increase exponentially as it has this century. We have examined the relative cost of electricity generated by coal and nuclear energy, particularly in those countries which are said to be critically dependent on Australian uranium. If coal is in the future made available to such countries at prices equivalent to recent prices, the cost of electricity generated from it is likely to be only marginally higher than the cost of electricity generated from nuclear energy.

We conclude that, while the economies of countries heavily dependent on imported oil have been adversely affected by increases in world oil prices, it is incorrect to say that there is a presently existing world energy crisis which will create disastrous economic effects. The nature and extent of the energy needs of countries which could conceivably be Australia's customers for uranium have been examined in Chapters 7 and 8, and it is clear that it is incorrect to suggest that there are energy impoverished nations which need Australian uranium for survival.

The major immediate world problem in the energy field is the availability of liquid fuels, and the provision of Australia's uranium will not do much to improve the situation. If Australia is to assist in the amelioration of this problem, concentration on the rapid development of alternatives to the world's diminishing petroleum resources appears desirable. Alternatives which have been suggested are the production of liquid fuels from coal and the provision of coal at economic prices to replace oil in existing and proposed electrical generating stations. In the longer term, further development of the technology to utilise solar energy in low and intermediate grade heat applications, in generating electricity and thence in the manufacture of liquid fuels such as methanol, appears desirable.

We are not suggesting that the existing nuclear industry be abandoned. Desirable as this may appear to be to some, it is clearly not practicable. We do conclude, however, that it is in Australia's best interests to encourage and assist the world community to do whatever is practicable to avoid, or where unavoidable to control, the development of inescapable reliance on nuclear power.

Put broadly, there are three lines of action which Australia could take which would assist in the achievement of this aim. They are the development of alternative energy sources and technologies, the conservation of energy, and the development of international control of nuclear technology. We suggest that the OECD, through its International Energy Agency (IEA), may prove to be the most suitable body through which these lines of action may be developed.

The Chairman of the Governing Board of the IEA, Ambassador Etienne Davignon, has described its aims, and the main points of an International Energy Program which it has developed. In order to reduce the dependence of participating countries on imported oil, co-operative programs are to be undertaken in the research and development of energy conservation, alternative energy sources and uranium enrichment.

We have said that Australia is shortly to become increasingly dependent on imported oil. It would benefit from co-operation with countries in a similar position and is well placed to assist in research and development of alternative energy sources. It possesses large coal resources and also receives immense quantities of unexploited solar energy. It has already developed some research and commercial competence in using solar energy for low grade heat.

Although Australia has for some years been an active member of OECD it is not a Member or a Participating Country in the International Energy Agency. We suggest that consideration should be given to Australia becoming a Member of the IEA.

So far as we are aware, Australia has not developed a National Energy Policy. We suggest that it should develop such a policy which should, in our view, encompass a National Energy Program giving emphasis to:

- research and development of energy conservation;
- education of the public in the need for energy conservation;
- the establishment of pricing, industrial and building policies which encourage energy conservation;
- research and development of alternative energy technologies, particularly the production of liquid fuels from coal;
- improvements in coal mining and burning technologies so as to reduce costs and environmental effects;
- the utilisation of solar energy.

Such a program, particularly if developed in co-operation with other members of the IEA, would help reduce the quantity and cost of Australia's future requirements for imported petroleum. Implementation of these proposals will also help to reduce the risks of proliferation.

Proliferation

The dangers of proliferation are very serious. They affect the whole world and exist whether or not Australia makes its uranium available to other countries.

The essential meaning of proliferation is the development in an increasing number of countries of nuclear weapons. The spread of technology and of nuclear power capacity give rise to fears of proliferation, but it is as well to separate cause from effect. One reason for this lies in practical considerations. It is unlikely that the spread of the peaceful uses of nuclear fission can be prevented, even if it is thought desirable to make the attempt. The dangers and difficulties associated with the production of nuclear power may lessen demand for it, but that is a different matter. It may not make very good sense for concerned nations to encourage positively the indiscriminate spread of peaceful programs, but the right of all countries to participate in the peaceful exploitation of the new energy source, if they wish to do so, can hardly be resisted. This was recognised when the NPT was being negotiated, and, as has been pointed out, that Treaty sought at the one time to prevent or discourage the spread of nuclear weapons and at the same time to encourage the spread of nuclear power for peaceful purposes. The Treaty would not have been made, and could not be maintained, in the absence of provisions giving non-nuclear weapon states every opportunity to develop nuclear power resources in their own countries.

The establishment of the IAEA and its safeguards system, designed to prevent the diversion of nuclear materials, equipment and information from peaceful to military purposes, long predated the NPT. Doubtless, when nuclear materials and technology were possessed by only two or three countries, this goal seemed a realistic one. Now there are five established nuclear weapon powers, India has exploded a nuclear device, and Israel either has nuclear weapons or, in the reported words of its President, 'if we need them, we will have them'. Four of these seven states are not parties to the NPT. What is of equal or greater significance is the number of additional states which now have advanced nuclear competence. Those, the nuclear weapon threshold states, will be able to develop nuclear weapons at relatively short notice. The additional factor, which is adding greatly to the current concern, is that some advanced nuclear states are selling to other states what is described as 'sensitive equipment'. This is the equipment which makes possible the production of the fissile material for atom bombs: separation, or reprocessing plants, from which plutonium can be obtained, and enrichment plants, which can be adapted for high enrichment.

It can be argued that the spread of nuclear weapons is not necessarily a bad thing, given that a number of states already have them. China takes the view that they should be in the hands of at least some socialist, non-imperialistic states, in order to counter the threat involved in their possession by a few 'imperialistic' states such as the U.S.A. and the U.S.S.R. The declared attitude of the great majority of nations is quite clear. It is expressed in two of the recitals to the NPT:

> 'Considering the devastation that would be visited upon all mankind by a nuclear war and the consequent need to make every effort to avert the danger of such a war and to take measures to safeguard the security of peoples,
>
> Believing that the proliferation of nuclear weapons would seriously enhance the danger of nuclear war, . . . '

Most people would endorse those views. The prospects of a world in which thirty, forty or more countries are equipped with nuclear weapons is terrifying. This is not to suggest that many countries will develop large arsenals of sophisticated nuclear weapons; only a few highly industrialised countries would have the resources to do so. What is a real possibility is that a number of countries, not being acknowledged nuclear powers, will make a small number of atom bombs, or explosive nuclear devices. They might acquire them by purchase, or theft, or by establishing nuclear facilities for the single purpose of making bombs, but these are in general unlikely methods. The more likely course is through a civil nuclear energy program.

It is thought that a state developing nuclear weapons in this way is much more likely to be doing so because of the effect their possession has in deterring aggression by others than because of any aggressive designs of that state. The acquiring state will probably be acting mainly if not solely out of fear for its own security. The trouble is that others will then fear it; and so proliferation will go on. Questions of international status or internal prestige may also be involved. By acquiring only a few atom bombs, a state might deter a near neighbour, and increase its bargaining strength in international power politics. The possession of large stocks of nuclear fission weapons by major powers, and in particular the super-powers, has so far brought about a stalemate, a condition of mutual deterrence. It is generally thought by strategists that this type of situation is less likely to be maintained with a larger number of states, each of which has nuclear weapons. Nor can the greatly superior nuclear might of the super-powers be relied upon to deter smaller states from using their nuclear weapons. The influence of the major powers is still great in many situations, but is thought by some to be less significant now than it was. Localised nuclear wars between medium or small states are conceived as quite possible. The consequences of such wars would be terrible enough, but the risk of escalation, with its catastrophic results, might be considerable.

It is probably inevitable if countries go on developing, or expanding, nuclear power programs that many will in due course have available to them the materials and technology which will enable them to make nuclear weapons themselves. This must always have been foreseen, but it does not mean that the future probability should be accepted as if it were present fact. There is solid advantage in gaining time; much can happen in the interval.

The present situation is giving great anxiety in a number of countries. In some, the need to take action has become a major political issue. The question of what to do has been the subject of full consideration in Committees of the United States Congress this year, was the subject of debate in the Canadian Parliament this year and is due to come before the United Nations soon. In January of this year, Secretary of State Kissinger repeated to a Senate Committee what he had said in September 1974 in a speech to the U.N. General Assembly:

> 'The world has grown so accustomed to the existence of nuclear weapons that it assumes they will never be used. But today, technology is rapidly expanding the number of nuclear weapons in the hands of major powers and threatens to put nuclear explosive technology at the disposal of an increasing number of other countries. I urge that immediate steps be taken to arrest this danger.'

The meeting of the London Suppliers' Club earlier this year was probably a manifestation of the general concern. Numbers of scientists with intimate knowledge of the scientific and technical considerations involved have this year expressed grave disquiet. Some are less concerned; some see the situation as already beyond control. A former chairman of the U.S. Atomic Energy Commission, Mr David Lilienthal, has repeatedly and as lately as this year drawn attention to the seriousness of the problem and exhorted his countrymen, and the nations of the world, to act before it is too late.

A full analysis of the position would involve considerations of a strategic and political nature which are beyond the scope of the present Inquiry. Although many views have been expressed to us, they mainly come from members of the public who have not made a close study of the international scene and are unacquainted with some of the material factors. The documentary evidence, on the other hand, contains discussions of the problem by informed people. While we are not in a position to make firm recommendations as to what course should be followed to lessen the danger, we are in a position to make some comments and some suggestions, and believe that we should do so.

Chapter 13 contains a discussion of limitations and weaknesses in the existing non-proliferation machinery, and suggestions for improvement are made in that discussion. We draw attention here to the list of suggestions made by the Congressional Research Service of the Library of Congress in a report prepared in March 1976 for the Committee on Government Operations of the United States Senate. The list is Appendix D to this Report.

The United States and the U.S.S.R. are both very concerned at the sale of sensitive equipment, separation plants and enrichment plants, by countries such as France and the Federal Republic of Germany. Undoubtedly, whatever safeguards agreements are insisted upon by those countries, the sales of the equipment, with the transfer of knowledge and skills which must sooner or later be involved, will increase the possibility that the countries acquiring the equipment will develop nuclear weapons. They may not, but it is highly likely that they will be able to do so if they wish. One of the purchaser countries is Pakistan, and it plainly faces the threat constituted by Indian nuclear weapons capacity.

A course offering promise of improvement lies in the direction of greater control of the manufacture, sale and use of sensitive equipment. It is convenient to refer again to some scientific and technical considerations. With some exceptions, separated plutonium is only at present necessary for military purposes, although it can be used as fuel in the types of reactors in use today. Enriched uranium is necessary for most reactors, but highly enriched uranium is at present, again with some exceptions, only necessary for making bombs. It is quite possible to say whether the nuclear reactors in operation in any country require separated plutonium, or highly enriched uranium.

It would be of considerable assistance to the non-proliferation objective if power reactors now supplied to countries not equipped with weapons did not require for their operations either plutonium (in separated form) or highly enriched uranium. If a country does happen to have reactors which require either of the materials mentioned, those materials would form part of made-up fuel elements. The country concerned might well be able to import the fuel elements rather than make them itself.

The critical importance of the separation (or reprocessing) operation in the case of plutonium and the high enrichment process in the case of uranium is obvious. Their importance is well recognised in the whole safeguards system. At present, and perhaps for a long time to come, the key to controlling the movement from civil to military nuclear capacity is to be found in those processes. It might therefore be possible to evolve a system to control proliferation which turns more directly than now is the case on the possession of the plant to separate plutonium or produce highly enriched uranium. The necessity will, of course, remain to control strictly the transfer of plutonium, at least when it is in separated form, and also of highly enriched uranium. The importance of the plant mentioned is heightened by the circumstance that nearly all countries on earth have resources of uranium in one form or another. In a country aspiring to make nuclear fission weapons, the source material will probably be at hand. The technology to produce the materials for a bomb, and to make a bomb, may be available or may be acquired by procuring the services of scientists and engineers from overseas, but in any event is not visible, and not readily controllable. The visible, tangible, element is the plant which is necessary to produce the fissile material.

A suggestion which has been made is to bring all such plant under international control, possibly by the IAEA, which has power under its constitution to assume that control. International ownership has also been mentioned. The granting to the owning or controlling body of 'extra-territorial rights', such as are now enjoyed by the IAEA, would be an important aspect. In line with that suggestion there has been active consideration for some time by the IAEA, and by organisations in the U.S.A., of the feasibility of establishing multinational fuel centres to enrich uranium and separate plutonium. They are envisaged as being subject to special safeguarding arrangements. A problem relating to these proposals is that, with relatively few enrichment and reprocessing plants, nuclear material, especially separated plutonium, may have to be in transit for greater distances, thus increasing the risks of theft, diversion and accident. The problem could be reduced if fuel fabrication plants were close to the separation and enrichment plants.

There are naturally a variety of difficulties. Quite a few countries now have plant of the nature in question, even though in some cases it is of a small, experimental or prototype nature. Some of these nations are not parties to the NPT. Then, there is the fact that the nuclear weapon states have, and will maintain the right to have, reprocessing or high enrichment capacity specifically in order to make weapons. They are unlikely to permit any form of international control or inspection of those activities.

Assuming that the necessary treaty arrangements could be made, there would be the risk of clandestine development of plant by some countries, contrary to treaty obligations. A country determined on such a course probably cannot be prevented in the long run, but a treaty system, relating to sale as well as manufacture and use, and combining inspection, might even in such a case delay development and might also give notoriety to it. The NPT and safeguards systems are themselves based on objectives not much less limited.

We have talked of dealings in nuclear equipment and materials as if they take place between countries. This is convenient, and is helpful in emphasising the strategic aspects of what is done. The fact is that transactions usually take place between corporations in the respective countries; for example, between large engineering and construction firms on the one hand, and power utilities on the other. Government control in some measure is likely to be present in all cases. It does seem, however, that excessive commercial enthusiasm in supplying countries has over the last decade played a major part in bringing about the present situation. The nuclear equipment supply business has become internationally competitive, and great sums of money are involved. Quite likely, the consequences could not

24647/76—**12**

be foreseen, but it is apparent to us that if nuclear proliferation is to be brought under control, so must the activities of suppliers of nuclear equipment and materials.

This seems to have been done already in the U.S.A. where Congress and the Executive, not to mention the regulatory bodies, now closely scrutinise major international sales and dealings in nuclear materials and equipment. It seems unfortunate, from the point of view of curbing proliferation, that approval was recently given to the sale of reactors by U.S. companies to Israel and Egypt, albeit subject to 'stringent safeguards'. Doubtless many considerations were involved, including the fact that President Nixon, influenced by the Middle East political situation, had promised Egypt a reactor or reactors several years before. The unfortunate aspect of the transactions is that neither country is a party to the NPT. A likely consequence of supplying them with reactors is to lessen the incentive of states without nuclear power to join the NPT, and some which are parties may become increasingly discontented. It is the fact that a high proportion of nuclear weapon threshold countries are not members of the NPT, having been influenced in their decisions, no doubt, by the consideration that they may wish to develop nuclear weapons. Something has to be done to bring them into the Treaty, or a revised Treaty.

We have already, in the early part of this Chapter, shown that the proliferation problem should not be seen in isolation from wider considerations. The desire for independent nuclear power capacity is related in many cases to the unequal global distribution of energy resources and, associated with that consideration, is the fact that some of the highly industrialised nations, and in particular the U.S.A., consume a greatly disproportionate part of those resources.

The spread of nuclear weaponry is likely to be the result of fear when confronted by neighbouring countries with such weapons. There is a need to give greater assurances of security to the countries which are or may be affected in this way. This is what they expect, and, as previously mentioned, something, but not much, has been done about it by the nuclear powers. Indeed, some of the lack of security is related, in some countries, to what is conceived as the possibility that the nuclear powers, whose development of weapons is protected by the NPT, will use their nuclear power as a threat against non-nuclear states in order to achieve their purposes. If a result is to be achieved, these aspects have to be acknowledged frankly and taken into consideration. Much could be solved if the super-powers, whose attitudes and actions provide the key to so much of the situation, could be persuaded to accept some degree of real arms limitation. The fact that the NPT is based on an acceptance of super-power supremacy would seem to carry a strong correlative obligation on the super-powers to make that Treaty work effectively.

Our final comment relates to the Organisation for Economic Co-operation and Development (OECD), which we have referred to in a number of places in this Report. It was established in 1960 and Australia joined it in 1971. The OECD has a Nuclear Energy Agency (NEA) which includes nearly all OECD members, and it is that agency which arranges and supervises the dumping of low level nuclear wastes from European countries in the Atlantic. It is concerned with the 'orderly development of nuclear energy for peaceful purposes'. We have mentioned the International Energy

Agency of the OECD. It was created in 1974, but Australia has not yet joined it. In the view of the Director of its Office of Long Term Co-operation, the OECD-IEA could play a part in promoting international collaboration on reactor licensing, waste disposal, financing, uranium supply and fuel services. The Organisation and its agencies are well regarded, and it occurs to us to suggest that they might provide avenues for initiatives designed to relieve the proliferation problem.

The general impression we have is that international tension in relation to the nuclear danger is increasing. Nevertheless, there are distinct rays of hope. There is already the remarkable situation that most of the states of the world are parties to the NPT, and all but three of these, being nuclear-weapon states, have renounced nuclear weapons. The other two nuclear-weapon states, France and China, are not parties to the Treaty. The most promising feature is that on this matter the U.S.A. and the U.S.S.R., and China also, have common cause. It can safely be assumed that none wants global nuclear war, and that they all see the possession by other nations of nuclear weapons as a threat to peace, and as a threat to themselves, having in mind in particular that nuclear war may escalate. Both the U.S.A. and the U.S.S.R. are members of the IAEA and parties to the NPT, and they came together recently as members of the London Club. Further, between them they currently export nearly all the enriched uranium which is exported in the world, the U.S.A. having the greater part of the market.

Australia occupies a very special position in the nuclear scene. It is represented on the Board of Governors of the IAEA, and was represented at the NPT Review Conference held last year. It is possessed of relatively large uranium reserves which by now have attracted world-wide attention. It has postponed developing these in order to inquire into the environmental consequences of doing so, involving all the global consequences. This also is widely known. It has a nuclear research establishment and a small nuclear industry of its own, but it is certainly not involved in the sale of equipment or technology associated with nuclear power generation. These features may combine, perhaps with others also, to put Australia in a favourable position to take initiatives, or at least to support them.

16 CONCLUSIONS

The issues The wider issues with which this Report deals arise out of grounds of opposition to the Ranger mining proposal. It was submitted that there were dangers associated with the various operations of the fuel cycle, from the mining of uranium to the production of power in reactors, that there were serious and unresolved problems concerning the disposal of radioactive wastes, that there were risks of terrorist theft and use of plutonium, and that there were increased risks of nuclear war flowing from nuclear proliferation. It was contended that the continuing development of the nuclear power industry would produce greater inequality between the developed and undeveloped countries, and that this, as well as being undesirable in itself, was likely to lead to increased international tension. It was submitted that, taken alone, some of those matters constituted sufficient ground for not mining, and that taken together they certainly did so. The central proposition was that, if Australia supplied its uranium to the industry, it would be contributing in some measure to each of those hazards and problems and that therefore it should not do so. To some extent, the argument rests simply on ethical values. In some important aspects, such as the dangers of high-level wastes, of terrorism and of proliferation, practical considerations affecting Australia arise. The submission was that mining should not take place at all, or should at least be postponed until it was clear that major problems, such as the disposal of wastes, had been overcome.

In further support of the submission, it was put that on economic grounds nuclear energy was not a satisfactory source of power, that it could only in any event offer a temporary way out of the energy problems of the countries wanting to use it, and that other sources of energy were preferable and could be developed. It was also submitted that nuclear power programs were less securely established than had been made to appear, and that there might well be a revulsion against them overseas. It was put that, for these and other reasons, the use of nuclear power would not develop as projected, with the consequence that there would be less demand for uranium and the profits would be less than predicted by the proponents and by others who support mining.

These were the principal matters relied upon. Others, not noted here, have been discussed in the chapters dealing with the topics to which they relate. Grounds of opposition which relate only to the particular Ranger proposal will be dealt with in the Commission's Second Report.

The submissions and arguments mentioned were countered by the proponents and by other witnesses. It was submitted that often the hazards were exaggerated by opponents of nuclear power, in some respects greatly so; that the economic and social suffering which would occur if nuclear energy were not developed would be greater than the hazards inherent in nuclear power; that the nuclear industry in all its aspects had to date a very good safety record, not least in relation to harm from radioactivity; that the hazards concerned had been exhaustively investigated by various authorities, were well understood and were under control; that the nuclear industries in countries likely to purchase our uranium were closely regulated and supervised; that the problem of high-level wastes had been virtually overcome by the proposal for vitrification and geological disposal; that the risk of terrorist activities was recognised and guarded against; that the safeguards systems provided sufficient protection against diversion and proliferation; that the operation of nuclear power stations was cleaner and involved less risk to people and the physical environment generally than fossil fuel stations; that a number of countries needed nuclear power, and a number had become dependent on it, at least in the short term; that the governments of many countries had accepted nuclear power, and it was not to the point, even if it were correct, to say that there was a large body of opposition to nuclear power development in their countries; that there was a considerable assured market for uranium; that (according to some witnesses) there was a risk that if permission to mine was not given soon, the market might shrink and prices drop because of the projected introduction of fast breeder reactors; and that the profits to be made were very good. It was submitted that, if Australia did not supply uranium, others would, and its abstention would make no difference in kind or degree to the presence of such hazards, difficulties and problems as there were.

An argument of a different kind relied upon by the parties opposed to mining was that if Australia were to decline to mine and sell its uranium specifically because of the hazards and problems involved, and were to announce its policy to the world, this would be likely to have an important effect in restricting further nuclear development, if not in actually causing a cut-back. The answer of the proponents, and others, was that such a course would be most unlikely to have the effect sought, but that, if it were desired to improve further the position in relation to the hazards and problems referred to, this could best be done if Australia were a supplier to the industry.

It was submitted by some witnesses that, whatever else happened, the Mary Kathleen mine should be allowed to continue to produce uranium to fulfil contracts already entered into. At the expected rate of production of about 1000 tonnes of yellowcake per annum, this would take about seven years. It was on the other hand contended by some that the contracts made before 1973 by Mary Kathleen Uranium Ltd, Queensland Mines Ltd and by Peko Mines Ltd and Electrolytic Zinc Company of Australasia Ltd should be honoured by using existing stocks at Lucas Heights, and, to the extent that they were insufficient, by making arrangements with other countries to provide uranium.

The proponents, and witnesses supporting their viewpoint, took the view in relation to some matters (not including, for example, proliferation) that such risks and problems as now exist are relatively minor, are of the order ordinarily accepted in everyday living, and will in all probability be overcome before they become at all acute or serious. It would be time enough to adopt a more draconian attitude if and when it was found that they were getting serious, and appeared intractable. Their opponents took more into account the long term future, as they saw it. They were of the view that humanity should not have to suffer added risks, even if they may not be great, and that the nuclear industry should be required to demonstrate that risks, particularly from radioactivity, were virtually negligible, before being allowed to develop any further. Associated with this viewpoint was the fear that if nuclear development was not stopped very soon, the industry would develop a momentum of its own, and be beyond effective control.

Some of the opponents placed reliance on a view that people in the developed countries should simplify their life-styles appreciably, so as to decrease the demand on non-renewable energy resources such as coal, oil and nuclear fuel. The scope for energy conservation, even with existing life-styles, was emphasised.

Assessing the evidence

The background of fact and opinion which is relied upon to give support to each of these submissions and arguments has already been examined, and in most cases the arguments themselves have also been discussed.

We have endeavoured to look at the issues, and examine the opposed contentions, in as objective a manner as possible. While the conclusions we arrive at must of necessity be our own, we have, where appropriate, endeavoured to apply the standards and values generally accepted in our society, as we understand them. To do otherwise would be to express purely personal opinions, and this would be both wrong and unhelpful.

It has to be acknowledged that certain standards and values accepted in the past and still used to interpret issues before us may not be held now by a substantial minority or even a majority of Australians. Attitudes on exploiting the nation's mineral resources are an instance where such changes may have occurred. Each Commissioner holds somewhat different values. We have given much thought to whether we should identify any varying interpretations we may hold. Our conclusion was that such an action would tend to be divisive in the ensuing debate through people supporting one or other of the expressed views rather than reviewing the issues for themselves.

Two further matters should be mentioned. There are present facts and future forecasts about the nuclear power industry. As Australia cannot supply uranium for at least three or four years and lead times for any development in the industry tend to be lengthy, it has been natural to look to the future to some extent. It seems to us that part of the polarisation of views which has occurred flows from the fact that, while the proponents place emphasis on the short and medium term, the opponents place considerable emphasis also on forecasts about the long term. They then bring to the fore problems which may arise if plutonium recycling is established and if the commercial fast breeder reactor becomes fact. We also are apprehensive about the possibility of such developments, but we think it unwise to base present decisions too much on forecasts of problems which may not occur. If at an appropriate time in the future it is found that they are likely to occur, their disadvantages, as then seen, will have to be weighed in the scales against the advantages of a continuing nuclear power industry.

The second matter is that rapid change is a feature affecting virtually every aspect of the nuclear industry. Whatever now appears to be the situation, it may appear very differently in even a few years. For this reason we will be recommending that, whatever course is now followed, there be instituted a system of regular review.

Different arguments of the opponents of mining have different logical bases, and they were not in all cases spelt out. In relation to those arguments, it is important to bear in mind that no intention has been announced to carry out in Australia any of the operations of the nuclear fuel cycle, except mining and milling. It is probable that for some time at least the nearest purchaser geographically will be Japan. No plan has been announced to establish any other operation of the nuclear power industry in Australia. If one is made at some future time, it will have to be evaluated separately. We were not told of any plan for Australia to receive nuclear wastes from any other country, and we assume there is none.

Risks in fuel cycle operations

It was not contended that, if properly regulated and controlled, hazards associated only with the mining and milling of uranium were of such a magnitude that those operations should not be allowed. There may nevertheless be quite natural concern that there may be a risk to health from releases of radioactivity in the course of those activities or after they have ceased. The topic has been dealt with in detail in Chapter 10. We are quite satisfied that, if properly regulated and controlled according to known standards, those operations do not constitute any health hazard which is greater in degree than those commonly accepted in everyday industrial activities.

It is convenient to deal next with the remaining operations of the nuclear fuel cycle, leaving aside for a moment the matter of wastes storage and disposal, acts of terrorists, and proliferation. The fact that the operations in question will take place in countries outside Australia is not, we agree, a ground for neglecting consideration of the hazards involved, but it must be an important matter to take into account. The fact is that most if not all the countries which are likely to be purchasing our uranium in the 1980s are advanced industrial countries. They include the U.S.A., Great Britain, the Federal Republic of Germany and Japan. The countries concerned all have well developed nuclear power industries and it can be assumed that they are well aware of the hazards and problems associated with the production of nuclear power. We appreciate the point that is made to the effect that in some countries secrecy surrounding the technical problems of the industry, and, more particularly, the occurrence of accidents, has deprived most members of the public of information which, had they known it, may have led to restrictions on further development, if not abandonment, of nuclear fission as a source of power. We are not able fully to evaluate this consideration. There is certainly evidence to suggest that nuclear power programs in the U.S.A. and Great Britain developed out of wartime nuclear weapons programs without the public having any real opportunity to scrutinise proposals and assess risks. As a very general observation, it can be said that in a number of the countries Australia might supply it is only recently that there has been an informed public awareness. We can take note of the existence of opposed points of view and the existence of substantial bodies of opposition to expansion of the industry, but we do not know the degree of opposition or the exact grounds of it, or even how fully or accurately informed it is. A factor which is bound to be of importance when people of any country weigh up the pros and cons is the degree to which, as they see it, their country is dependent on nuclear energy. The fact is that in most of the countries which will be seeking supplies of Australian uranium, nuclear reactors have been part of the scene for some years, and several more years must elapse before those countries can receive supplies from us. If in that time the hazards are found to be unacceptable in those countries, they can be expected to take appropriate action. This would probably be by way of restricting or reducing nuclear power production, with consequent effect on requirements for uranium.

Leaving aside questions respecting terrorism, nuclear wastes and proliferation, our assessment of the position is that, while the operations of the nuclear power industry need close regulation and constant surveillance, they probably do not entail risks greater in sum than those inherent in alternative energy industries. Certainly those risks provide no proper basis for a refusal on our part to supply the advanced industrial countries which are likely to be our customers. It is possible that, in the future, the operations of a nuclear power industry in a particular country, or the operation of those industries generally, will be found to be more hazardous than now appears. This could simply be the consequence of growth, but is more likely to be the result of information coming to hand as the result of further research and experience. A consequence might then be that Australia would decide, in the interests of that country, that it should not supply. A more likely consequence is that the country or countries concerned will either not seek any supplies or will only seek reduced supplies. The situation is apt to become self-adjusting.

Radioactive wastes

The matter of wastes involves some different considerations. Low-level and intermediate-level wastes are being disposed of in ways that seem to us to be reasonably satisfactory from the point of view of people in the countries concerned, and constitute no problem likely to affect Australia. Nevertheless, like all aspects of nuclear power production, the position requires constant watching. Some wastes are already being dumped in the ocean, albeit under close supervision. They do not at present constitute a problem, but could do so if supervision were relaxed, or if the operation became too widespread, or the bulk too great.

High-level wastes constitute a serious potential problem. Because of the long life of some of the isotopes, and the extraordinarily long life of a few, the problem must be viewed more in a global context than is the case with other aspects of the nuclear fuel cycle. It is almost universally agreed by governments and by the nuclear industry that those wastes must be disposed of in such a way that they will remain isolated from the biosphere for hundreds of thousands of years. While experiments have been going on for many years, no method of disposal which clearly will satisfy this requirement has been proved. Many scientists are satisfied that vitrification and geological disposal is a safe method, but this is still in the experimental stage.

Used fuel elements are highly radioactive and contain what, after chemical separation, will be the high-level wastes. At present, except where reprocessing is carried out as part of the regular operation of the fuel cycle, the fuel rods are stored in water tanks or ponds. It was thought that all these fuel elements would in due course be reprocessed to yield their constituents, such as plutonium, uranium, and fission products, but there is now a question in some countries about the extent to which that will be done. In any event, there is the problem of what is eventually to happen to the long-lived radioactive constituents.

Some wastes have already been separated out, and these are stored in liquid form in steel tanks. While storage is reasonably satisfactory, provided care is taken and the ponds or tanks are kept in a secure location under regular observation, it can only be an interim measure. No country which Australia is likely to supply has attempted to dispose of high level nuclear wastes permanently. The problem is not at present regarded as acute, partly because of expectations about the feasibility of existing proposals for permanent disposal, and partly because the quantity is not yet so large that the long-lived highly radioactive materials could not if necessary be disposed of by known means with small risk of damage to the environment.

While we do not think that the waste situation is at present such as to justify Australia wholly refusing to export uranium, it is plain that the situation demands careful watching, and, depending on developments, regular and frequent reassessment. If, even in a few years, satisfactory disposal methods have not been established, it may well be that supplies of uranium by Australia should be restricted, or even terminated. This would in any ordinary case only be done after consultation with the purchasing country, which would probably also be the country where the wastes would be and which would therefore be more directly affected by the problem.

We also suggest that, whether or not Australia supplies uranium, it endeavours to have some internationally acceptable system established for the disposal of high-level wastes, and international supervision of what is done. In this connection it is to be remembered that a large amount of high-level waste has already been generated by the nuclear weapons industries of nuclear weapons states; and more will be produced. We do not know what has happened to the wastes produced in the U.S.S.R., but obviously the whole problem is one of first-rate international importance. The creation anywhere of large amounts of long-lived radioactivity is a matter of concern to all people, because of the very serious global effects which dissemination of them could have.

The terrorist threat

Terrorism is a real danger, and one likely to have international consequences. Whether terrorists would make a bomb and actually use it seems doubtful at present, even though that appears technically and organisationally feasible in the light of current terrorist capabilities. The continued spread of nuclear technology must increase the opportunities for and the probability of such actions. While provision of security adequate to guarantee against terrorist intrusion is theoretically possible, even as against the highly competent terrorist and criminal organisations which now exist, there must be a question whether adequate precautions will in fact be taken. The possible development of terrorist organisations as de facto military units operating on behalf of a hostile but unidentified country compounds the difficulty. The problem, common to all considerations of nuclear technology, is that, while the risk of the event may be very small, the consequences can be shockingly great. In our view the possibility of nuclear terrorism merits energetic consideration and action at the international level. We do not believe that this risk alone constitutes a sufficient reason for Australia declining to supply uranium. It does however provide a further reason why the export of our uranium, including what is proposed to be done with it, and where, are matters which the Government should keep under constant scrutiny and control.

Proliferation

The most serious danger in our view is that of proliferation of nuclear weapons. This has already been discussed at length in Chapters 12, 13 and 15. As mentioned in Chapter 15, the problem exists whether or not Australia supplies uranium. There are then two quite separate considerations. One is whether by supplying its customers Australia will be contributing to the danger that proliferation presents. The other is whether, having the resources, it can adopt some course such as the refusal or postponement of supply which might encourage the nations of the world to act to diminish the threat.

Whether the situation will be improved or made worse if Australia supplies uranium will depend on whom it supplies and on what terms. It could supply a country which is about to embark on a nuclear weapons program. Then it may be contributing, in greater or less degree, but nevertheless directly, to an increase in the danger of nuclear war. It would not do so knowingly because that would be contrary to its obligations under the NPT. In our view it should not supply countries which are not parties to the NPT. The purchasing countries, if not nuclear weapons countries, will be obliged under the NPT not to use the uranium for explosive devices. Stringent safeguards provisions in sale agreements would also assist in minimising the risk that any uranium exported by us to them will be used for any military purpose. Two of the countries we are likely to supply are nuclear weapons powers already. Selling them uranium would not be likely to increase proliferation, even if they were to use it for military purposes. Resale by them (and any other buyers) without consent should be provided against. It is possible that considerations of our own defence might in any event outweigh any factors adverse to the supply to those countries of our uranium.

It may be that by supplying some countries we would help to relieve those pressures which can lead to armed conflict, nuclear or non-nuclear. In relation to some countries, especially Japan, there are very important mutual trade considerations which could make very undesirable a decision never to supply. A total refusal to supply uranium could incline some countries, of which Japan might again be a leading example, to turn earlier than would otherwise be the case to the less uranium-hungry fast breeder reactor, assuming it to be commercially successful. We are in any event faced with our own Treaty obligations and in particular Article IV of the NPT. This Article has been discussed. It does not impose legally enforceable obligations, and current practice in some states is not to let it interfere unduly with strategical defence and economic considerations. It is nevertheless part of the spirit as well as the letter of the Treaty that we should assist peaceful nuclear programs. Our recommendation is that if uranium is to be exported this be done selectively. The criteria of selection need no elaboration, but flexibility will also be necessary, because the scene will be an ever-changing one.

Permanent refusal to supply

We mentioned earlier an argument that Australia should permanently refuse to supply uranium, or should at least postpone supply, with a view to persuading other countries, by our example, from entering upon or further developing nuclear power production. Although the argument probably finds its strongest support from considerations of proliferation, it can be supported by reference to all the hazards and problems of the industry.

A total renunciation of intention to supply designed to bring an end to all nuclear power industries or all further development of them would in our view be likely to fail totally in its purpose. If the purpose were simply to draw international attention to the dangers of and associated with the industry, that purpose might be achieved, but it is most unlikely that any worthwhile action would result. On the other hand there are positive reasons against adopting such a course. Apart from financial considerations, which are not to be neglected, there are considerations to which we referred when dealing with the topic of proliferation. A total refusal to supply would place Australia in clear breach of Article IV of the NPT and could adversely affect its relation to countries which are parties to the NPT. These matters might not have been of any concern at all had we not advanced our preparations for uranium mining to the stage they have now reached, so that our readiness and ability to supply within a few years are now obvious. We are of the view that total renunciation of intention to supply is undesirable.

Postponement of supply

When it comes to a question of a temporary postponement of supply, considerations intrude which are different and more varied. The need to take steps to impede proliferation is the matter of most serious concern. We cannot be sure that an announced intention not to mine or export for a period of, say, two to five years, will not have an impact leading to more vigorous international action than might otherwise take place. Nor can we be sure of the extent of any adverse international reaction, having in mind, in particular, our NPT obligations and our relations with our close trading and strategic partners. Apart from the question whether action such as that under consideration would be successful, or is on balance desirable, there is the different consideration whether during such a delay some of the existing problems may be more satisfactorily resolved. Initiatives are in progress or can be taken which offer some chances of success in improving safeguards and helping to prevent diversion of nuclear material for war-like purposes. Australia, as a country which had no stated intention to withhold permanently its uranium might be able to exert influence to improve matters through channels we have mentioned. There is a possibility that during this period technological advances will reduce hazards to man and the environment, especially with regard to the treatment and disposal of high-level radioactive wastes. On the evidence available to us no country with an expressed intention to buy Australian uranium will in the meantime be dependent on Australia, in the sense that supplies at reasonable cost could not be obtained elsewhere. Japan is perhaps the country most likely to need Australian uranium and it has already contracted for supply of all its requirements until 1985, almost entirely with countries other than Australia. When taken in conjunction it seems to us that these factors make delay an option which might reasonably be followed.

The contrary argument contends that any delay would serve no useful purpose and mining should be permitted provided the recommended controls, regulations and conditions are put into effect immediately. The case for immediate development rests mainly on the positive advantages from predicted economic benefits and also on the view that Australia, as an active supplier of uranium, would be better placed to exert its influence towards reducing the problems associated with a nuclear industry. Millions of dollars have already been invested in preparatory work for uranium mines in Australia and returns on this substantial investment would be postponed by a delay. Additional costs would be involved in interest on borrowed capital, maintenance of work already completed and retention of a core of personnel for established proposals. The assessment of any losses in national income which would be caused by a delay is highly speculative. However, a two year delay in the whole uranium industry, assuming it would otherwise have developed at the highest foreseeable rate, as discussed in Chapter 9, could cause losses to national income of some tens of millions of dollars. On the other hand, if a delay caused an increase in world uranium prices or if the assumed maximum rate of development was not achievable, losses could be small, even negligible. We recognise the importance of these factors and would not contemplate suggesting that a delay be considered if we were not convinced that the hazards associated with the nuclear industry are of overriding national and international significance.

The essential difference between the bases for these two alternative propositions is that the case for delay places greater weight on the considerations of the environmental and human costs of nuclear development whereas the case for immediate, albeit cautious and restrained, commencement of an Australian uranium export industry places greater weight on assumed economic benefits and the view that, as a result of commencing exports, Australia's influence will be correspondingly greater.

Because the evidence from which each line of argument is derived remains conjectural and also for reasons stated earlier when discussing the proliferation problem, we have not found a compelling basis for a conclusion on the question whether it is preferable to delay coming to a decision about mining for a period of several years or alternatively to proceed with carefully planned development of the industry. What we do conclude is that at present Australia should not commit itself to withholding for all time its uranium supplies, and that it should take the course which is determined to be the most effective and most practical in order to bring a favourable response from other states in relation to the proliferation problem.

Regulation and control

A number of countries which are involved in producing uranium for the nuclear power industry have by legislation given strong controls in respect of that activity to the central government. By way of illustration, reference may be made to the Canadian Atomic Energy Control Act, Section 9; the Indian Atomic Energy Act, 1962, Sections 3, 4 and 5; and the South African Atomic Energy Act, 1976, Sections 5 and 7. In some countries legislative measures are designed, in part, to give effect to treaty obligations.

Reference has already been made to the powers given the Commonwealth under the *Atomic Energy Act* 1953. It may be doubted whether a power relative to mining in the states which is limited legislatively to reliance on 'defence purposes' is adequate. On the contrary, it is likely to be productive of uncertainty and disputation. There would seem to be little doubt that the Commonwealth has constitutional power to assume full control for all proper purposes, and we suggest that consideration should be given to its doing so. The *Customs Act* 1901 and Regulations thereunder already give power in respect of the import of radioactive substances and export of uranium.

It will already be apparent that we are strongly of the view that, if the mining and selling of uranium proceeds, it should be on a strictly controlled and regulated basis. We see uranium as a highly strategic material, the supply of which involves not only questions concerning the hazards and problems we have been discussing, and trade considerations, but also foreign policy and defence considerations. In our view it cannot be treated as an ordinary trade commodity, or even an ordinary fuel commodity. The purpose of the controls and regulations should be as follows:

(a) To ensure that supplies can be stopped, or restricted, if the hazards and other difficulties and problems to which we have referred make such a course desirable. In particular, the Government should be unfettered in adopting a course which promises to ease the proliferation problem. It may not be a sufficient objection to the adoption of such a course to say that Australia's supply of uranium was not making matters worse, or even that others might supply if we do not. We recognise that political, economic, and social pressures may combine to render very difficult the suspension either temporarily or permanently of a flourishing and profitable uranium mining industry. Factors that would make this difficult include waste of capital investment, international problems in relation to countries which have become dependent on Australian uranium, probable instability of the world's uranium market following Australia's withdrawal, and disruption of an established work force.

(b) To enable discretion to be exercised in the selection of the countries to be supplied and in the extent to which they will be supplied. We recognise that the exercise of such a discretion may create problems in international relations, but we see no escape from the need to exercise it if occasion should arise. We are clearly of the view that no country which is not party to the NPT should be supplied.

(c) To help ensure market stability. The market is a very sensitive one, which is likely to be subject to major fluctuations. The maintenance of any sort of production and price stability will be very difficult.

(d) To ensure the orderly and economic development of mining activities. It follows from all that we have said that this is necessary. In our view, the situation must be scrupulously avoided in which industry pressure based on existing investment or commitment is allowed to be a factor in determining the course to be followed in relation to the more serious of the hazards mentioned. It seems to us that the controls we have in mind would best be achieved by a sequential development of mines, but this is not a matter upon which we wish to express a concluded opinion in this Report.

(e) To ensure that the most satisfactory possible safeguards arrangements are made, and that they fully satisfy our treaty obligations.

(f) To maintain all such controls as are possible to ensure that resale by a purchaser of Australian yellowcake only takes place with Australia's approval.

(g) To enable the Government to be satisfied in advance that the conversion and enrichment of Australian uranium will not create serious hazards, particularly of diversion or proliferation.

(h) To ensure that there can be a ready response to Parliamentary decisions. We believe that regular and frequent consideration of the problems by Parliament is quite essential.

Some purposes which relate more directly to national considerations will be dealt with in the Second Report.

We appreciate fully that the institution of a system of controls such as we have mentioned will present difficulties to the mining enterprises, and may increase costs and reduce sales and profits. In particular, planning beyond a relatively short period of time will not be easy. We see no escape from those disadvantages, serious though some of them may be, if Australia is to participate in the nuclear power industry with a due sense of responsibility to the people of the world and its own people.

One of the arguments which has been used against any mining development is that, once it is started, no government will have the strength to resist the pressures for its continuance and even its expansion. We believe that this is a serious consideration. If the argument is a sound one, then our proper course is to recommend against commencement. What seems to us to be vital is that a clear and firm statement of policy be made at the outset, supported, from the outset, by a strong system of control. If this is done, it is our hope and expectation that governments will be able to maintain effective control for the purposes we have mentioned. One way in which the position of government may be weakened is if foreign corporations with large interests in other countries obtain control of the management of Australian mining operations. The present Government's policy of limiting foreign ownership and maintaining Australian control is of first importance in this regard. It should however be recognised that managerial control can readily arise from ownership by or on behalf of the one corporation of a relatively small proportion of share capital.

Uranium Advisory Council

There are three further matters, upon which we have already touched, which relate to the question of government policies and government control. The first is that the Government will require an advisory body to assist it, particularly in relation to environmental aspects such as those dealt with in this Report. We recommend the formation of what might be called the Uranium Advisory Council. We see it as important that this body command public confidence. To this end and so as to widen its own appreciation of relevant problems, the public should have access to it. As well as advising the Government, it should report to Parliament. It should have a reasonably large membership, and should include a number of people who are not expert in nuclear science or technology. The majority of its members should not be involved in the nuclear industry or the promotion of nuclear power. It should have a substantial degree of independence, and should be able to insist on production to it, from whatever source, of relevant materials. Its functions should in substance be limited to questions affecting the exploitation of Australian uranium in relation to the nuclear power industry, although it should be able to consider some incidental questions, such as alternative energy sources. It would be important that it should be able to advise upon the adequacy of safeguards and their application to countries to which it is intended to supply Australian uranium. The Council could, by common membership or otherwise, have close association with a national energy council. Some further delineation of the role and constitution of this Advisory Council may be appropriate when we deal specifically with matters in the Northern Territory.

The second matter is related to the first. It is quite clear that in Australia and overseas there are great, and growing, numbers of people who are very concerned about the development of the nuclear power industry. It seems to us to be vital, particularly to those who wish to see the continuance of the industry, that the public be kept fully informed of relevant facts. We have already noted that there is a tendency on the part of some to misrepresent those facts. One way in which accurate information can be supplied is through the activities of an Advisory Council such as we have mentioned. Another is by reasonably frequent parliamentary debate, not only in the Federal parliament, but in State parliaments and Territory legislative bodies. We strongly urge that steps be taken to ensure that such debates take place.

Review

The third of the three matters to which we have referred is the need for periodical review of policies. We have already said that the basis of present views can be altered or completely changed, within a matter of a few years, as the result of further research or experience. We recommend that a system of periodical review be established. This should be done by a body reasonably independent of Government and to which the public has access. It may prove to be a task that could be carried out by or under the aegis of the Advisory Council.

Plutonium and the fast breeder reactor

In this Report we have discussed in some detail the serious dangers associated with the recycling of plutonium but have not said much concerning the fast breeder reactor. This is largely because it is not likely to play a major role this century. It is nevertheless seen by many in the nuclear industry as the logical next step after the introduction of commercial recycling of plutonium in today's thermal reactors. It is feared by some that, without it, uranium supplies will not sustain the nuclear fission industry for very long. Its advent will be a matter of international importance for a number of reasons, but particularly because it will make unavoidable the development of what has been called the plutonium economy. While this development would also follow the introduction of large-scale recycling in thermal reactors, there is evidence that reprocessing of oxide fuels may not be economically justified, and may not proceed unless required by governments. Some of the opposition to the proposal to mine and sell Australian uranium has been based more on the spectre that the future introduction of a plutonium economy is seen to present than on problems with the existing industry. We have deep reserve about its introduction, for reasons we have already canvassed. Our recommendation is that the development of plutonium recycling and the fast breeder reactor be kept under close and critical observation so that Australia may take such action, internationally or otherwise, as seems most appropriate to counter the hazards they seem certain to present.

The Northern Territory

It has been submitted to us that there are some factors, peculiar to the position in the uranium province in the Northern Territory, that reinforce conclusions applicable generally to the uranium mining industry, and which, when taken with other matters, should lead to a decision not to develop the mines in question at all. Those submissions remain for consideration.

PRINCIPAL FINDINGS AND RECOMMENDATIONS

These findings and recommendations are to be read and understood in the context of the Report as a whole and with particular reference to the sections of the Report in which they are respectively discussed.

1. The hazards of mining and milling uranium, if those activities are properly regulated and controlled, are not such as to justify a decision not to develop Australian uranium mines.

2. The hazards involved in the ordinary operations of nuclear power reactors, if those operations are properly regulated and controlled, are not such as to justify a decision not to mine and sell Australian uranium.

3. The nuclear power industry is unintentionally contributing to an increased risk of nuclear war. This is the most serious hazard associated with the industry. Complete evaluation of the extent of the risk and assessment of what course should be followed to reduce it involve matters of national security and international relations which are beyond the ambit of the Inquiry. We suggest that the questions involved are of such importance that they be resolved by Parliament. In Chapters 15 and 16 we have gone as far as the terms of reference and the evidence permit in examining the courses open and in making suggestions.

4. Any development of Australian uranium mines should be strictly regulated and controlled, for the purposes mentioned in Chapter 16.

5. Any decision about mining for uranium in the Northern Territory should be postponed until the Second Report of this Commission is presented.

6. A decision to mine and sell uranium should not be made unless the Commonwealth Government ensures that the Commonwealth can at any time, on the basis of considerations of the nature discussed in this Report, immediately terminate those activities, permanently, indefinitely or for a specified period.

7. Policy respecting Australian uranium exports, for the time being at least, should be based on a full recognition of the hazards, dangers and problems of and associated with the production of nuclear energy, and should therefore seek to limit or restrict expansion of that production.

8. No sales of Australian uranium should take place to any country not party to the NPT. Export should be subject to the fullest and most effective safeguards agreements, and be supported by fully adequate back-up agreements applying to the entire civil nuclear industry in the country supplied. Australia should work towards the adoption of this policy by other suppliers.

24647/76—**13**

9. A permanent Uranium Advisory Council, to include adequate representation of the people, should be established immediately to advise the Government, but with a duty also to report at least annually to the Parliament, with regard to the export and use of Australian uranium, having in mind in particular the hazards, dangers and problems of and associated with the production of nuclear energy.

10. The Government should immediately explore what steps it can take to assist in reducing the hazards, dangers and problems of and associated with the production of nuclear energy.

11. Policy with regard to the export of uranium should be the subject of regular review.

12. A national energy policy should be developed and reviewed regularly.

13. Steps should be taken immediately to institute full and energetic programs of research and development into (a) liquid fuels to replace petroleum and (b) energy sources other than fossil fuels and nuclear fission.

14. A program of energy conservation should be instituted nationally.

15. The policy of the Government should take into account the importance to Australia, and the countries of the world, of the position of developing countries concerning energy needs and resources.

Our *final recommendation* takes account of what we understand to be the policy of the Act under which the Inquiry was instituted. It is simply that there should be ample time for public consideration of this Report, and for debate upon it. We therefore recommend that no decision be taken in relation to the foregoing matters until a reasonable time has elapsed and there has been an opportunity for the usual democratic processes to function, including, in this respect, parliamentary debate.

POSTSCRIPT:

The Sixth Report of the British Royal Commission on Environmental Pollution

We refer in this Report to the United Kingdom Royal Commission on Environmental Pollution which, under the Chairmanship of Sir Brian Flowers, has conducted an inquiry into environmental risks from the nuclear industry. The Report of the Royal Commission, which has the title Nuclear Power and the Environment, was presented to the U.K. Parliament on 22 September 1976. Copies reached us after our own report had been written, but we add this postscript for the purpose of drawing attention to the very valuable discussion which the U.K. Report contains, and to some of its principal recommendations.

Britain already has a large nuclear industry, based on gas-cooled reactors and including all stages of the nuclear fuel cycle except mining and milling. The Royal Commission was only incidentally concerned with problems arising in other countries and in this respect the scope of its Inquiry was somewhat narrower than ours. Its Report did not deal in detail with the problem of proliferation, although on this topic the Commission does say:

> '. . . the spread of nuclear power will inevitably facilitate the spread of the ability to make nuclear weapons and, we fear, the construction of these weapons'. (p. 76)

and

> 'Indeed, we see no reason to trust in the stability of any nation of any political persuasion for centuries ahead. The proliferation problem is very serious and it will not go away by refusing to acknowledge it'. (p. 76)

Some issues, such as reactor safety and waste disposal, are dealt with more thoroughly in the British Report than has been possible in the present Report. Given these differences in emphasis, the views of the British Royal Commission are, we believe, in general agreement with the views of this Inquiry, so far as they relate to the same matters.

Some of the principal conclusions of the Royal Commission are as follows:

With respect to reactor safety: 'The risk of serious accident in any single reactor is extremely small; the hazards posed by reactor accidents are not unique in scale nor of such a kind as to suggest that nuclear power should be abandoned for this reason alone.'

With respect to the security of plutonium: 'Plutonium appears to offer unique potential for threat and blackmail against society because of its great radiotoxicity and its fissile properties.'

With respect to radioactive wastes: 'There should be no commitment to a large programme of nuclear fission power until it has been demonstrated beyond reasonable doubt that a method exists to ensure the safe containment of long-lived, highly radioactive waste for the indefinite future.'

With respect to general nuclear policy for the U.K.: 'The dangers of the creation of plutonium in large quantities in conditions of increased world unrest are genuine and serious. We should not rely for energy supply on a process that produces such a hazardous substance as plutonium unless there is no reasonable alternative.'

'The abandonment of nuclear fission power would, however, be neither wise nor justified. But a major commitment to fission power and a plutonium economy should be postponed for as long as possible.'

'There should be increased support for the development of other energy sources including energy conservation, combined heat and power systems and fusion power.'

APPENDIXES

APPENDIX A: LIST OF WITNESSES

Mr A. W. Ackhurst
Mr P. D. Adamson
Mr K. F. Alder
Mr H. R. Allen
Miss S. H. Allen
Mr C. C. Allom
Mr R. G. Anderson
Mr W. J. Arnott
Dr J. Arriens
Mr D. A. Atkinson
Mr P. J. Bailey
Mr P. Balminidbal
Mr P. H. Barratt
Mr N. C. Barrett
Dr M. Barson
Sir Philip Baxter
Mr L. J. J. Beens
Mr A. I. Bellingham
Mr D. K. Bellingham
Dr C. H. Berndt
Prof. R. M. Berndt
Mr F. L. Bett
Mr P. J. Bills
Mr Yorky Billy
Mr A. D. Blain
Prof. J. O'M. Bockris
Mr R. C. Borton
Mr P. G. Braham
Prof. D. F. Branagan
Mr H. Brennan
Mr G. Briscoe
Mr T. J. Brooks
Mr G. Brown
Mr Isiah Buranali
Sir Macfarlane Burnet
Mr O. J. Cameron
Dr J. A. Camilleri
Mr I. Campbell
Mr R. K. Carruthers
Mr G. Chaloupka
Mr B. M. Chapman
Prof. W. W. S. Charters
Rev. B. A. Clark
Mr G. H. Clark
Mr D. E. Clarke
Mr J. E. V. Collins
Sir Willis Connolly
Mr J. E. Cook
Mr T. C. Cooke
Dr H. C. Coombs
Dr J. M. Costello
Dr A. J. R. Cotter
Dr J. R. Coulter
Dr K. A. W. Crook
Mr D. J. Crossley
Rev. R. L. Croxford
Mr T. C. Dalby
Mr L. K. Dalton
Mrs J. L. Darling
Prof. H. L. Davies
Dr P. S. Davis
Mr D. R. Davy
Mr H. J. de Bruin
Dr D. Doley
Mrs A. Dunlop
Mr H. J. Dunster
Mr Raymond J. Edwards
Mr Robert Edwards
Dr J. H. Eedle
Mr H. F. Eggington
Rev. A. F. Ellemor
Mrs M. Elliott
Mr G. S. Eupene
Mr E. C. Evans
Dr J. E. Falk
Mr T. O. Fegan
Prof. P. J. Fensham
Mr K. E. Ferrier
Mr M. R. Finger
Mr W. J. Fisher
Dr M. E. Flood
Dr T. M. Florence
Mr L. J. Forrester
Dr H. Frith
Mr R. M. Fry
Mr T. Gangali
Mr P. E. J. Gardner
Mr N. C. Gare
Mr P. L. Garton
Mr A. D. Gaulton

Mr R. J. Giles
Mr G. W. Godwin
Chief Inspector K. P. Grant
Mr S. L. Grassie
Prof. H. S. Green
Dr J. G. Greenhill
Mr R. J. Gregory
Mr A. J. Grey
Mr R. H. Gruber
Dr F. W. Gunz
Mr J. F. Halfpenny
Dr J. Y. Hancock
Prof. S. F. Harris
Mr F. R. Hartley
Mr J. P. Hauser
Mr D. R. Hawke
Mr P. J. Hayes
Mr L. C. Helberg
Mr I. D. Henderson
Mrs A. Heuzenroeder
Mr C. T. Hignett
Dr D. J. Higson
Mr D. G. Hill
Mr S. J. Hodge
Mr E. R. Hollamby
Mrs M. J. Holmes
Mrs S. Holmes
Prof. Z. J. Holy
Mr A. D. L. Hooper
Mr D. R. Hughes
Mr J. C. Hunter
Dr P. L. T. Ilbery
Mr M. F. Ivory
Dr C. Jack-Hinton
Mr M. B. Jacob
Mrs R. Jaquier
Mr J. Kamminga
Mr E. Kaptein
Mr I. Keen
Mr W. J. Kelty
Mr L. G. Kemeny
Mr R. J. Kentish
Miss S. L. Kerrigan
Mr B. Knight
Mr H. Kurth
Mr J. S. Lake
Mr Jambana Lalara
Mr Jambani Lalara
Father J. J. Lanigan
Dr R. E. Lapp
Mr J. T. Larkin
Miss J. Law
Mr T. R. Lawler
Mrs A. D. Lawrie
Prof. G. W. Leeper
Mr P. E. Leske
Dr G. A. Letts
Mr W. Lichacz
Mr D. A. Lindner
Dr C. R. Lloyd
Mr J. C. Lough
Mr T. C. Lovegrove
Mr A. B. Lovins
Mr A. G. McArthur
Mr R. J. McCabe
Mr D. McColl
Mr J. C. D. McDonnell
Mrs V. McGregor
Mr A. H. McIntosh
Dr N. J. Mackay
Dr D. G. MacPhee
Mr B. McPhie
Mrs R. G. Maralngurra
Mr S. Maralngurra
Mr D. Marshall
Mr P. W. Marshall
Mr D. F. Martin
Miss J. C. Miller
Mr P. A. Miller
Mrs J. Minnards
Dr R. H. Mole
Mr J. F. Moloney
Mr G. W. Montgomerie
Mr R. E. Morrison
Dr J. G. Mosley
Mrs A. Moyal
Mr Dheimbalipu Munungurr
Mr S. P. Myers
Mr R. Nazer
Mr R. S. Needham
Mr W. Neilley
Mr J. H. Nicholls
Mr L. T. Nicholls
Mr N. F. Nunan
Mr K. P. O'Connor
Mr P. J. O'Connor
Mr R. F. O'Donoghue
Dr R. J. O'Neill
Mr M. D. Pardee
Mr A. A. Parker
Mr J. D. Pendarvis
Dr R. Pengilley
Dr N. Petersen
Mr R. Phelps

Mr C. L. Pierrehumbert
Dr F. K. Pittman
Dr A. B. Pittock
Dr J. H. Price
Mr A. B. Pulsford
Mr P. Purich
Mr I. R. Radford
Prof. B. J. F. Ralph
Mr T. D. Redhead
Mr H. Reed
Dr P. R. Reeves
Rev. S. A. Reid
Dr J. M. Rendel
Mr L. Reynolds
Mr W. J. Ricketts
Dr A. P. Roberts
Mr S. Roberts
Mr D. J. Robertson
Mr F. P. J. Robotham
Mr D. J. Rochford
Mr R. Ross
Miss M. L. Rowe
Mr B. H. Ruse
Mr V. S. Russell
Mr G. R. Ryan
Mr W. M. Scriven
Mr L. S. Shea
Mr J. P. Shelton
Dr A. Sibatani
Mr A. H. Silvester
Dr R. J. Sinclair
Mrs M. Smith
Mr W. G. Smith
Prof. L. E. Smythe
Prof. R. L. Specht
Dr P. H. Springell
Mrs V. Stanton
Mr D. J. Stephens
Mr M. N. Stephens
Mr E. D. J. Stewart
Mr P. R. Stork
Mr H. Stretton
Mr Strider
Mr K. D. Suter
Mr A. C. Svenson
Mr D. E. Symon
Dr J. L. Symonds
Mr S. R. Tanner
Mr J. L. O'N. Tedder
Mrs I. T. I. Temple
Prof. R. W. Thompkins
Mr A. L. Tidswell
Sir Ernest Titterton
Mr J. R. Tomlinson
Mr G. R. Trefry
Mr A. S. Trippe
Mr N. Tsatsaronis
Mr M. J. Tuck
Mr N. Turner
Mr J. A. Waddams
Mr S. Wagbara
Mr K. S. Wallis
Mrs M. E. Walsh
Miss B. K. Ward
Mr W. R. Watson
Mr G. Weissman
The Hon. W. C. Wentworth
Mr R. N. Wesley-Smith
Mr J. F. B. Whelard
Miss D. White
Mr J. A. A. Wilders
Mr R. C. Wilkinson
Mr A. R. Williams
Dr C. H. Williams
Miss H. Williams
Dr N. J. Williams
Mr A. F. Wilson
Mr H. O. Wilson
Mr F. Woerle
Mr D. T. Woods
Mr W. J. K. Wright
Mr R. J. Wyatt
Mr E. Wyndham
Mr B. F. Wynn
Mr J. A. Yeates
Mr D. G. D. Yencken
Rev. Dr A. Yule

24647/76—**14**

APPENDIX B

Nuclear power reactors around the world at 31 December 1975. Units smaller than 30 MWe are omitted

	Operating			Being Built		On Order		Total		
	Number	*Capacity (GWe)*	*Percentage*	*Number*	*Capacity (GWe)*	*Number*	*Capacity (GWe)*	*Number*	*Capacity (GWe)*	*Percentage*
United States of America	53	36.4	50.6	62	62.8	99	111.1	214	210.3	55.6
Canada	6	2.5	3.4	6	4.1	8	5.2	20	11.8	3.1
Japan	11	5.9	8.2	12	9.3	1	0.3	24	15.5	4.1
Belgium	3	1.7	2.4	..	..	4	3.8	7	5.5	1.5
Federal Republic of Germany	7	3.3	4.6	11	10.2	8	9.8	26	23.3	6.1
France	10	2.8	3.9	16	14.7	6	5.5	32	23.0	6.1
Italy	4	1.4	1.9	..	..	4	3.9	8	5.3	1.4
Luxembourg	..	..	..	..	..	1	1.3	1	1.3	0.4
Netherlands	2	0.5	0.7	..	..	..	..	2	0.5	0.1
United Kingdom	28	5.3	7.4	11	6.5	..	..	39	11.8	3.1
Total EEC	54	15.0	20.9	38	31.4	23	24.3	115	70.7	18.7
Austria	..	..	..	1	0.7	..	..	1	0.7	0.2
Finland	..	..	..	3	1.5	1	0.7	4	2.2	0.6
Spain	3	1.1	1.5	7	6.3	2	1.9	12	9.3	2.5
Sweden	4	2.4	3.3	6	4.9	1	1.0	11	8.3	2.2
Switzerland	3	1.0	1.4	2	1.9	3	3.0	8	5.9	1.6
Yugoslavia	..	..	..	..	..	1	0.6	1	0.6	0.2
Total Other Western Europe	10	4.5	6.2	19	15.3	10	7.2	37	27.0	7.3
Argentina	1	0.3	0.4	1	0.6	..	..	2	0.9	0.2
Brazil	..	..	..	1	0.6	2	2.6	3	3.2	0.9
India	3	0.6	0.8	3	0.6	2	0.4	8	1.6	0.4
Iran	..	..	..	..	..	4	4.2	4	4.2	1.1
South Korea	..	..	..	1	0.6	2	1.2	3	1.8	0.5
Mexico	..	..	..	2	1.3	..	..	2	1.3	0.3
Pakistan	1	0.1	0.2	..	..	..	..	1	0.1	0.03
Philippines	..	..	..	..	..	2	1.3	2	1.3	0.3
Taiwan	..	..	..	4	3.1	2	1.8	6	4.9	1.3
Total Developing Countries	5	1.0	1.4	12	6.8	14	11.5	31	19.3	5.0
Bulgaria	2	0.9	1.3	2	0.9	..	..	4	1.8	0.5
Czechoslovakia	1	0.1	0.2	3	1.3	1	0.4	5	1.8	0.5

German Democratic Republic	3	1.0	1.4	2	0.9	2	0.9	7	2.8	0.7
Hungary	..	..	..	2	0.9	2	0.9	4	1.8	0.5
Poland	..	..	..	..	..	1	0.4	1	0.4	0.1
Romania	..	..	..	..	..	1	0.4	1	0.4	0.1
U.S.S.R.	12	4.6	6.4	8	5.9	5	3.9	25	14.4	3.8
Total Countries with Centrally Planned Economies	18	6.6	9.3	17	9.9	12	6.9	47	23.4	6.2
World Total	157	71.9	100.0	166	139.6	165	166.5	488	378.0	100.0

Source: *Nuclear News*, World List of Nuclear Power Plants, February 1976.

APPENDIX C

A. STATES WHICH ARE PARTY TO THE NPT, AS OF 31 AUGUST 1976

Afghanistan
Australia
Austria
Bahamas
Belgium
Bolivia
Botswana
Bulgaria
Burundi
Cameroon
Canada
Central African Republic
Chad
Costa Rica
Cyprus
Czechoslovakia
Dahomey (now Benin)
Denmark
Dominican Republic
Ecuador
El Salvador
Ethiopia
Fiji
Finland
Gabon
Gambia
German Democratic Republic
Germany, Federal Republic of
Ghana
Greece
Grenada
Guatemala
Guinea-Bissau
Haiti
Holy See
Honduras
Hungary
Iceland
Iran
Iraq
Ireland
Italy
Ivory Coast
Jamaica
Japan
Jordan
Kenya
Khmer Republic[1] (formerly Cambodia, now Democratic Kampuchea)
Korea, South
Laos[1]
Lebanon
Lesotho
Liberia
Libyan Arab Republic
Luxembourg
Madagascar (Malagasy Republic)
Malaysia
Maldives
Mali
Malta
Mauritius
Mexico
Mongolia
Morocco
Nepal
Netherlands
Netherlands Antilles[2]
New Zealand
Nicaragua
Nigeria
Norway
Paraguay
Papua-New Guinea[3]
Peru
Philippines
Poland
Romania
Rwanda
San Marino
Senegal
Seychelles
Sierra Leone
Singapore
Somali Democratic Republic
Sudan
Surinam
Swaziland
Sweden
Syrian Arab Republic
Taiwan[4]
Thailand
Togo
Tonga
Tunisia
Upper Volta
Uraguay
Union of Soviet Socialist Republics
United Kingdom of Great Britain and Northern Ireland
United States of America
Venezuela
Vietnam, Republic of[5]
Western Samoa
Yemen Arab Republic
Yugoslavia
Zaire

B. STATES, MEMBERS OF THE UNITED NATIONS, NOT PARTY TO NPT, AS OF 31 AUGUST 1976

Albania
Algeria
Argentina
Bahrain
Bangladesh
Barbados
Bhutan
Brazil
Byelorussian Soviet Socialist Republic
Cape Verde
Chile
China, People's Republic of
Colombia
Comoros
Congo
Cuba
Egypt, Arab Republic of
Equatorial Guinea
France
Guinea
Guyana
India
Indonesia
Israel
Kuwait
Malawi
Mauritania
Mozambique
Niger
Oman
Pakistan
Panama
People's Democratic Republic of Yemen
Portugal[6]
Qatar
Sao Tome and Principe
Saudi Arabia
South Africa
Spain
Sri Lanka
Trinidad and Tobago
Turkey
Uganda
Ukrainian Soviet Socialist Republic
United Arab Emirates
United Republic of Tanzania
Zambia

C. OTHER STATES NOT PARTY TO THE NPT, AS OF 31 AUGUST 1976

Angola
Dubai
Liechtenstein
Nauru
Switzerland[7]

These lists were compiled from the most recent information available to the Commission.

Notes:

[1] By a generally accepted principle of international law, a state, which undergoes a change of government, is bound by treaties entered into on its behalf by the previous government. However, it has long been part of the Communist view of international law that this principle has no application to a situation where the change of government is effected by a 'revolution of the people' which creates a new Communist state quite distinct from the former state which it replaces. Hence, while the two Western depositary governments, the United Kingdom and the United States, regard Laos and Cambodia as parties to the NPT, it is possible that neither of the latter two states regards itself as a party to treaties entered into by the former non-Communist governments of those states; and in the absence of any communication from these states, it is doubtful whether they can be considered as parties to the NPT.

[2] A party to the NPT by virtue of the Netherlands' ratification of the NPT.

[3] Became a party to the NPT by virtue of Australia's ratification. In the absence of a communication to the contrary from this new state, the depositary governments may consider it, at this stage, as a party to the NPT.

[4] Became party to the NPT as the Republic of China.

[5] The de facto division of North and South Vietnam was brought about by the Geneva Conference of 1954 and prolonged by the failure of the states concerned to implement the requirements in the Accords that free elections be held as a step towards reunification of the country. At that time, Western countries recognised the Saigon government as the government of the whole of the Republic of Vietnam, whereas the Soviet Union, the People's Republic of China, and other Communist states recognised the government of Hanoi as the government of (the Socialist

Republic of) Vietnam. The ratification of the NPT was the act of the Saigon Government on behalf of the Republic of Vietnam. In the light of the Communist victory in the Vietnam War and the recent measures takes to unify the country as the Socialist Republic of Vietnam, it is doubtful whether Vietnam can be regarded as a party to the NPT, although it is believed that the United Kingdom and the United States are continuing to list Vietnam as a party which has ratified the Treaty.

[6] It is understood that on 2 July 1976 the President of Portugal approved Portugal's accession to the NPT and the relevant decree (No. 588/76) became effective on 22 July 1976. Information is not available as to whether Portugal's instrument of accession has been deposited with the depositary governments.

[7] Ratification of Switzerland's signature of the NPT has been approved by the upper house of the Swiss Parliament; the matter has now been referred to the lower house. A final decision on ratification is not expected before December 1976.

APPENDIX D

List of Possible Means to Limit Proliferation

I. *Reducing pressures for proliferation*

- Convince non-weapons nations that nuclear weapons are not necessary for their security
- Step up nuclear disarmament and arms control. Divert weapons material to power production
- Conclude an effective comprehensive nuclear test ban treaty
- Cause more nations to ratify the NPT and forswear nuclear weapons
- Improve ways for nations to settle disputes without resort to threats or use of nuclear force
- Provide assurance against undetected development of nuclear weapons
- Establish enforceable sanctions against nations that prepare to or begin to make nuclear weapons
- Get nuclear weapons nations to pledge no first use of nuclear weapons against non-weapons states

II. *Limiting the spread of the ability to make nuclear weapons*

- Encourage nuclear interdependence to prevent proliferation of capabilities to make fissionable materials
- Establish multinational, regional nuclear fuel cycle centres
- Have nuclear supplier nations offer reliable nuclear fuel services
- Establish a cartel of nuclear supplier nations to restrict the export of nuclear fuel cycle technology and products to NPT nations and to sell nuclear fuels and fuel services at prices below those national enterprises would have to charge
- Re-establish a U.S. dominance in the world nuclear fuel cycle market by expanding facilities and charging prices low enough to discourage national fuel cycle ventures
- Speed up and increase development of nuclear fuel cycles that do not contain materials readily usable for nuclear explosives

III. *Making theft or diversion more difficult*

- Substantially increase physical security for nuclear materials and critical facilities
- Establish an international security force for nuclear materials both within nations and at multinational centers
- Establish national and international means to anticipate and intercept attempted theft or diversion, and to pursue and recover stolen materials
- Establish theft of nuclear materials as an international crime
- Limit nuclear fuel exports to nations to normal or low enriched uranium. Prohibit their use of plutonium, uranium-233 or highly enriched uranium
- Store, rather than reprocess, nuclear fuels, so that plutonium will not be recovered. Pay non-weapons nations for fuel value of plutonium and unrecovered uranium in their fuel

– Limit use of breeder reactors to weapons nations or to operation by multinational organisations

IV. *Making more certain the detection and announcement of theft or diversion*

– Improve national and international safeguards systems and technologies to promptly and reliably detect and announce theft or diversion
– Openly publish results of safeguards audits and inspections

V. *Developing world alternatives to nuclear energy*

– Create economically feasible alternatives to expanded use of nuclear energy including solar energy, wind power, and fusion
– Offer technical assistance in non-nuclear energy only to nations that adhere to the NPT

This list was included by the Congressional Research Service of the Library of Congress in a report which it prepared in March 1976 for the Committee on Government Operations of the U.S. Senate.

GLOSSARY

Accession: A means by which a state, not having signed an international treaty, expresses its consent to be bound by the treaty. In the case of a multilateral treaty, accession is normally effected by the deposit, with the depositary government or governments named in the treaty, of a document called an 'instrument of accession'.

Accounting Procedures: A term used in the context of safeguards to mean the procedures by which a record is kept of all movements of nuclear materials.

Actinides: Elements with 89 or more protons in their nucleus. They include uranium and plutonium. Many are long lived alpha emitters.

Alpha particle: A heavy particle produced by a radioactive decay process and in various nuclear reactions. It consists of two protons and two neutrons and thus carries two positive charges. It is identical with the nucleus of a helium atom.

Background level: The naturally occurring level of radiation to which all people are exposed.

Beta particle: A light particle produced in many nuclear reactions and in radioactive decay processes. It may carry a negative or positive charge, but in common usage the term refers to the negatively charged particles which are identical to electrons.

Breeder reactor: A type of reactor in which the number of fissile nuclei produced (bred) from fertile nuclides is greater than the number of fissile nuclei concurrently destroyed. The type of breeder reactor currently being developed produces plutonium-239 from uranium-238, while consuming plutonium-239 and uranium-235.

Burn-up: A measure of the quantity of energy that has been obtained from a sample of nuclear fuel in a reactor core. It also determines the quantity of fission products produced and hence influences the level of radioactivity of the spent fuel. Burn-up is usually measured in units of megawatt-days per tonne of fuel.

Capacity factor: The ratio of the amount of electricity actually generated by a power station in a given period to the amount of electricity which would have been generated if the power station had been operating at full capacity throughout the period.

Centrally planned economies: A grouping of countries used by the United Nations for many purposes including the compilation of world energy statistics. Countries with centrally planned economies are the U.S.S.R., the countries of eastern Europe, China, Mongolia, North Korea and Vietnam.

Cladding: The hollow sheath inside which nuclear fuel is hermetically sealed to make a fuel rod. Cladding may be magnox, stainless steel, zirconium alloy (zircaloy), or ceramic.

Containment: The structures, within and including the reactor building, which are designed to prevent any material which may escape from the reactor itself reaching the outside environment. The reactor containment usually consists of steel and thick concrete.

Control rods: Rods, containing a material that strongly absorbs neutrons, designed to be inserted into a reactor core to control the rate of the nuclear reaction.

Conversion: In the nuclear fuel cycle, the stage at which yellowcake (impure uranium oxide) is processed to a composition and purity suitable for fuel fabrication or for isotopic enrichment.

Cooling ponds: In the nuclear fuel cycle, the water-filled tanks at a power station or reprocessing plant in which spent fuel rods are placed after removal from the reactor.

Core: The central part of a nuclear reactor containing the fuel rods, moderator and control rods. The nuclear fission reactions take place and the resultant heat is generated within the core.

Critical: Of an assembly of nuclear materials, fulfilling the condition that it is just capable of sustaining a nuclear chain reaction.

Critical mass: The minimum mass of fissile material which can be made critical under given conditions with a specified geometrical arrangement and material composition.

Curie: The unit of radioactivity, defined as 37 000 million (3.7×10^{10}) disintegrations per second. This is almost equal to the radioactivity of 1 gram of radium-226. The abbreviation for curie is Ci. Quantities of radioactive material are commonly measured in curies; one curie of material is the quantity having an activity of one curie.

Decay product: The substance formed by radioactive decay of a radioactive nuclide. A substance, such as uranium-238, which decays through a sequence of steps has associated with it many successive decay products.

Depleted uranium: Uranium in which the content of the fissile isotope uranium-235 is less than the 0.71 per cent normally found in natural uranium.

Deuterium: A heavy isotope of hydrogen, hydrogen-2, in which the nucleus contains one proton and one neutron. The nucleus of normal hydrogen contains one proton only. Deuterium is not radioactive and occurs on earth as a small proportion of total hydrogen (see *Heavy water*).

Developed countries: A grouping of countries used by the United Nations for many purposes including the compilation of world energy statistics. They include the U.S.A., Canada, all the countries of western Europe including Yugoslavia and Greece, Israel, South Africa, Japan, Australia and New Zealand.

Developing countries: All the countries of the world other than the developed countries and countries with centrally planned economies. This group includes nearly all the poorer countries of the world; it includes all the countries of Central and South America, and most of the countries of Africa and Asia.

Dose: The amount of energy delivered to a mass of material by ionising radiation passing through it.

Dose-rate: The time rate at which ionising radiation delivers energy to a mass of material through which it is passing.

Energy analysis: A method of examining economic activities and products which seeks to estimate their cost in units of energy rather than in the monetary units which economists normally use.

Enriched uranium: Uranium in which the content of the fissile isotope, uranium-235, is higher than the 0.71 per cent normally found in nature. Low enriched uranium, containing 2 to 4 per cent of uranium-235, is used as fuel in many types of reactor. High enriched uranium, which may contain more than 90 per cent uranium-235, is used as fuel in some types of reactor and also to make nuclear weapons.

Enrichment: The process in the nuclear fuel cycle in which enriched uranium is produced.

External cost: A cost to society as a whole which is not generally reflected in financial transactions.

External dose: A dose of radiation delivered to a person (or other living organism) from outside the body, as opposed to internal dose delivered by radioactive materials that have been swallowed or otherwise taken into the body.

Fast reactor: A type of nuclear reactor in which the concentration of fissile nuclei in the fuel is so high that the nuclear reaction can be sustained by fast (i.e. unmoderated) neutrons.

Fault-tree analysis: A method of examining complex engineering systems to determine the probable occurrence and consequences of an accident in any component of the total system. It involves tracing the sequences of events in all parts of the system that could result from any particular accident. It was used in the *Reactor Safety Study*.

Fertile: As applied to a nuclide, it means capable of absorbing (capturing) a neutron and being thereby subsequently transformed into a fissile nuclide. Uranium-238 and thorium-232 are important fertile isotopes, giving rise to plutonium-239 and uranium-233 respectively.

Fissile: Capable of undergoing fission.

Fission: The process by which a nucleus splits into two approximately equal fragments plus several free neutrons, giving off large amounts of energy which appears as the energy of gamma radiation and heat. Fission occurs spontaneously in certain heavy elements, but the kind of fission making chain reactions possible occurs when a fissile nucleus absorbs a neutron.

Fission products: The mix of nuclides resulting from fission. Fission products are often unstable and undergo radioactive decay, emitting radiation of a range of types and energies.

Fossil fuel: Coal, oil, natural gas or other carbon-containing fuels derived from the fossilisation of living matter (mostly plants) that flourished hundreds of millions of years ago.

Fuel rod or pin: The form in which nuclear fuel is used in most types of reactor, consisting of a single tube of cladding filled with pellets of fuel. Fuel rods are made in a fuel fabrication plant.

Fusion: Short for thermonuclear fusion, a type of nuclear reaction in which two light nuclei, such as deuterium, fuse together to form one heavier nucleus, in the process releasing a large amount of energy. Fusion reactions only take place at exceedingly high temperatures. They are the source of the energy given off by the sun and also of most of the energy released when a hydrogen bomb explodes.

Gamma radiation: A form of electro-magnetic radiation, similar to light or X-rays, distinguished by its high energy and penetrating power. Gamma radiation is emitted from many nuclei when undergoing radioactive decay and in many other nuclear reactions.

Geothermal energy: Energy obtained, in the form of heat, from the depths of the earth. The most convenient form of geothermal energy is steam emitted from the surface of the earth in areas of volcanic activity, but geothermal energy is also contained in hot rock underground in many localities.

Gonad: Sexual or reproductive organ of an animal or plant.

GWe: Abbreviation for 'gigawatt electrical'. One gigawatt is equal to one-thousand megawatts electrical (1000 MWe). See also *MWe*.

Half-life: The period of time in which half the nuclei in a given sample of a particular radioactive nuclide will undergo decay. The half-life is a characteristic of a particular nuclide.

Heavy water: Water in which the hydrogen atoms are the heavy hydrogen isotope, deuterium. It is sometimes called deuterium oxide and occurs in natural water to the extent of about 1 part in 6000.

Hex: Short for uranium hexafluoride, UF_6, the gaseous compound of uranium which is used for enrichment by gaseous diffusion and centrifugation.

High level waste: The most highly radioactive waste from fuel reprocessing, containing most of the fission products from spent fuel and typically containing millions of curies per cubic metre when first separated. It also contains small amounts of unseparated uranium and plutonium, plus the greater proportion of the other actinides produced in the reactor.

Induced radioactivity: The radioactivity of nuclides produced from naturally stable nuclides, as the result of nuclear reactions with neutrons or high energy particles. Nuclear reactions between neutrons and some of the nuclei of the stable materials, for example in a reactor core structure, result in the formation of radioactive nuclides, some with rather long half-lives.

Intermediate level waste: A somewhat arbitrary classification of part of the waste from fuel reprocessing, typically containing thousands of curies per cubic metre.

Internal dose: See external dose.

Ionising radiation: Radiation which, by reason of its nature and energy, interacts with matter to remove electrons from (ionise) the atoms of material absorbing it, producing electrically charged atoms which are called ions.

Isotope: Atoms of an element having the same number of protons in their nuclei but different numbers of neutrons are called isotopes. All isotopes of an element have the same chemical properties and thus cannot be separated by chemical means. However, they can be separated by using certain physical processes, such as gaseous diffusion.

Kilotonne: One-thousand tonnes.

Kilowatt: A unit of power, the rate of using or producing energy. One kilowatt is equal to about one and a third horsepower.

Light water: Ordinary water with normal hydrogen atoms, as distinct from heavy water.

Low level waste: Part of the waste from various stages of the nuclear fuel cycle typically containing a few curies per cubic metre.

Magnox: A type of magnesium alloy used as cladding for the metallic fuel in early British and French gas-cooled reactors. In Britain, the word has been adopted as the name of the reactor type.

Material balance area: A term used under the safeguards accounting system to mean an area, which may include a nuclear fuel cycle facility, or several facilities or part of a facility, such that the area can be treated as self-contained for material accounting purposes. The quantity of nuclear material in each transfer into or out of the area can be measured and the inventory of material in the area can also be measured.

Millirem: One-thousandth of a rem.

Moderator: A material used in a reactor core to slow down fast neutrons, without unduly absorbing them, so as to increase the probability of the neutrons causing fission in a uranium-235 or plutonium-239 nucleus.

MWe, MW(th): One megawatt (MW) is a unit of power equal to one thousand kilowatts. MW(th) denotes the thermal power of a power station, that is the rate at which heat is produced (by fission in the reactor core if it is a nuclear power station). MWe denotes the electrical power output of the station and is only a fraction of the thermal power—typically about 33 per cent for a light water reactor and up to 40 per cent for a modern fossil fuel-fired power station. This ratio is called the thermal efficiency of the power station.

Neutron: An uncharged particle which is a constituent of the nucleus of all nuclides except hydrogen; neutrons are ejected from the nucleus in some types of nuclear reaction, including fission.

Nuclear material: Term used in the safeguards system to cover any special fissionable material or source material.

Nuclear reaction: Any event involving a change in the nucleus of an atom.

Nuclide: A nuclear species, all the atoms of which contain similar nuclei.

OECD: The Organisation for Economic Co-operation and Development with headquarters in Paris. Its member states are: Australia, Austria, Belgium, Canada, Denmark, Federal Republic of Germany, Finland, France, Greece, Iceland, Ireland, Italy, Japan, Luxembourg, Netherlands, New Zealand, Norway, Portugal, Spain, Sweden, Switzerland, Turkey, U.K. and U.S.A.

Opportunity cost: The value of the opportunity foregone by not using a resource in ways other than the way it is actually used.

Pressure tube: In some types of nuclear reactor such as CANDU and SGHWR, the pressurised coolant flows through a series of tubes, called pressure tubes, each of which contains a bundle of fuel rods.

Pressure vessel: In most types of reactor, including LWRs and gas cooled reactors, the pressurised coolant is confined in a single large vessel made of welded steel or concrete. This is called a pressure vessel, and the whole of the reactor core is contained within it.

Proton: A positively charged particle which is a constituent of the nucleus of all nuclides. The number of protons in the nucleus determines the chemical properties of an element and hence is characteristic of each of the chemical elements

Rad: The unit used to measure the energy of radiation absorbed by matter (the radiation dose); one rad is equal to the absorption of 100 ergs of the energy carried by the radiation per gram of material.

Ratification: A means by which a state, on whose behalf an international treaty has been signed, expresses its consent to be bound by the treaty. In the case of a multilateral treaty, ratification is normally effected by the deposit, with the depositary government or governments named in the treaty, of a document entitled an 'instrument of ratification'.

Reactor: A nuclear reactor is a structure within which a fission chain reaction can take place under controlled conditions; in a power reactor the energy of the chain reaction is harnessed to produce steam for electricity generation.

Recycling: The stage in the nuclear fuel cycle, not yet achieved on a commercial basis, in which uranium and plutonium extracted from spent fuel are returned to an earlier stage of the fuel cycle for reuse in fresh fuel rods.

Rem: A unit of radiation dose-equivalent, that is radiation dose compensated to allow for the greater relative damage to biological material caused by some particles, such as alpha particles and fast neutrons, and to compensate for other conditions of irradiation. For most beta and gamma radiation, one rem is equivalent to one rad of absorbed radiation dose.

Reprocessing: The stage of the nuclear fuel cycle, not yet achieved on a commercial basis expect for magnox fuel, at which plutonium and uranium in spent fuel are separated from the other actinides and the fission products, which constitute waste.

Reserves: As applied to a mineral, the quantity of a material, known to be present following geological exploration, which can be economically recovered with existing technology.

Resources: As applied to a mineral, the total quantity estimated to be present, including the quantity not yet established but expected to be present in existing proven areas, and speculations about other potential discoveries.

Safeguards: A series of international arrangements designed to detect and hence deter the use of nuclear facilities or materials for prohibited purposes, such as the production of nuclear weapons.

Signature: Signature of an international treaty on behalf of a state may merely be a means by which the text of the treaty is established as authentic and definitive. Signature of a treaty may, however, be the means by which a state expresses its consent to be bound by the treaty, if the treaty so provides, or if the states which negotiated the treaty so agree, or if that is the expressed intention of the state on whose behalf the treaty is signed. Some treaties require that signature be followed by ratification.

Source material: A term used in the context of safeguards to mean any material from which special fissionable material may be produced, including natural uranium, depleted uranium or thorium.

Special fissionable material: A term used in the context of safeguards to mean any material containing uranium enriched in uranium-235 or in uranium-233, or containing plutonium.

Tailings: The waste material from a uranium mill after the uranium has been extracted from the ore. Tailings contain the radioactive decay products of uranium mixed with a large volume of non-radioactive rock, all in a finely ground form and mixed with water.

Tails: The depleted uranium produced at an enrichment plant.

Tails assay: The proportion of uranium-235 remaining in the tails after enrichment. It is typically between 0.2 and 0.3 per cent. The tails assay is an important characteristic of the operation of an enrichment plant; the higher the tails assay, the less will be the amount of enrichment energy required to produce a given quantity of enriched uranium, but the larger will be the amount of natural uranium needed to produce that quantity of the desired product.

Thermal: Of neutrons, that they are travelling at a relatively slow speed, comparable with that of gas molecules at ordinary temperatures.

Thermal reactor: A type of nuclear reactor in which most of the nuclear fissions are caused by thermal neutrons that have been slowed down by a moderator. All power reactors being commercially installed at present are thermal reactors. They consume more fissile material in total, in the form of uranium-235 and plutonium-239 bred in the reactor, than they produce in the form of plutonium-239.

Tritium: A radioactive isotope of hydrogen with a nucleus containing one proton and two neutrons. It occurs naturally in minute quantities, but is produced as a by-product of controlled nuclear fission. It is also produced, in relatively large quantities, by nuclear fusion.

U_3O_8: The formula for uranium oxide in the form in which it is contained in yellowcake. Use of the formula gives an indication of the quantity of uranium in a given amount of the oxide.

Working Level: The quantity of radon decay products in one litre of air which will result in the ultimate emission by them of 130 000 million electron volts of alpha particle energy. If the short-lived decay products are in equilibrium with the radon in air, then 100×10^{-12} curies per litre of radon is equivalent to one Working Level.

Working Level Month: One Working Level Month is a measure of the total radiation dose that would be received by someone breathing air containing radon decay products at a concentration of 1 Working Level throughout the working period of a month (defined as 170 hours).

Yellowcake: The mixture of uranium oxides and impurities (typically about 95 per cent U_3O_8) produced at a uranium mill.

Yield: In the context of nuclear weapons, yield is a measure of the energy output of an explosion. It is usually expressed in terms of the amount of TNT which would produce the same explosive effect.

ABBREVIATIONS

AAEC	Australian Atomic Energy Commission
AGR	Advanced Gas-Cooled Reactor
AUPF	Australian Uranium Producers Forum
BWR	Boiling Water Reactor
CANDU	Canadian Deuterium Uranium Reactor, a type of PHWR
ECCS	Emergency Core Cooling System
EEC	European Economic Community
EIS	Environmental Impact Statement
ERDA	United States Energy Research & Development Administration
FBR	Fast Breeder Reactor
FEA	United States Federal Energy Administration
GCFBR	Gas-Cooled Fast Breeder Reactor
HTGR	High Temperature Gas-Cooled Reactor
IAEA	International Atomic Energy Agency
ICRP	International Commission for Radiological Protection
LMFBR	Liquid Metal Cooled Fast Breeder Reactor
LWR	Light Water Reactor
NPT	Treaty on the Non-Proliferation of Nuclear Weapons
NRC	United States Nuclear Regulatory Commission
OECD	Organisation for Economic Co-operation and Development
OECD–IEA	International Energy Agency of the OECD
OECD–NEA	Nuclear Energy Agency of the OECD
OPEC	Organisation of Petroleum Exporting Countries
PHWR	Pressurised Heavy Water Reactor
PWR	Pressurised Water Reactor
SGHWR	Steam Generating Heavy Water Reactor

24647/76 R76/150 Cat. No. 76 3287 4